This vibrant collection provides a nuanced, multifaceted exploration of the intricacies of the temporal and the global, especially in relation to human imagination and sociality. A must for scholars from across the disciplines interested in processes and practices connected to globalization.

Imre Szeman, *Canada Research Chair in Cultural Studies, University of Alberta, Canada*

Because it opens up and deepens the ways we can think critically of time and temporality – and shows why doing so is more urgent than ever – this volume is essential reading.

Robert Latham, *York University, Toronto, Canada*

Time, Globalization and Human Experience

This edited volume focuses on the intersection of time and globalization, as manifested across a variety of economic, political, cultural, and environmental contexts. Since David Harvey's influential characterization of globalization as "time-space compression," ample research has looked at the spatial aspect of the phenomenon, yet few have focused on globalization's *temporal* aspects. Meanwhile, other publications have analyzed problems of speed, acceleration, and the commodification of time, but while it often serves as the implicit or explicit backdrop for these studies of time, globalization is not investigated as a problem or a question in its own right. In response, this volume develops these conversations to consider how time shapes globalization, and how globalization affects our experience of time.

The interplay between varying aspects of the human experiences of time and globalization requires the type of interdisciplinary approach that this volume takes. The contributors advance an understanding of global time(s) as an arena of contestation, with social, political, ecological, and cultural implications for human and other lives. In considering the diverse valences of time and globalization, they illuminate problems as well as possibilities. Topics covered include emerging infectious diseases, temporal sovereignty, worker exploitation and resistance, chronobiology, energy politics, activism and hope, and literary and cinematic representations of counter-temporalities, offering a rich and varied account of global times.

This volume will be of great interest to students and researchers from a range of disciplines, including anthropology, cultural studies, globalization, international relations, literary studies, political science, social theory, and sociology.

Paul Huebener is an Assistant Professor of English in the Centre for Humanities at Athabasca University, Canada.

Susie O'Brien is an Associate Professor in the Department of English and Cultural Studies at McMaster University, Canada.

Tony Porter is a Professor of Political Science at McMaster University, Canada.

Liam Stockdale is a Postdoctoral Fellow at the Paul R. MacPherson Institute for Leadership, Innovation and Excellence in Teaching at McMaster University, Canada.

Yanqiu Rachel Zhou is an Associate Professor at the Institute on Globalization and the Human Condition and the School of Social Work, McMaster University, Canada.

Rethinking Globalizations

Edited by Barry K. Gills, University of Helsinki, Finland and Kevin Gray University of Sussex, UK

This series is designed to break new ground in the literature on globalization and its academic and popular understanding. Rather than perpetuating or simply reacting to the economic understanding of globalization, this series seeks to capture the term and broaden its meaning to encompass a wide range of issues and disciplines and convey a sense of alternative possibilities for the future.

Full series title: www.routledge.com/Rethinking-Globalizations/book-series/RG.
Most recent titles:

62. **The Redesign of the Global Financial Architecture**
    The return of state authority
    Stuart P. M. Mackintosh

63. **Markets and Development**
    Civil society, citizens and the politics of neoliberalism
    Edited by Toby Carroll and Darryl S. L. Jarvis

64. **Occupying Subjectivity**
    Being and becoming radical in the 21st century
    Edited by Chris Rossdale

65. **Globalization and Global Citizenship**
    Interdisciplinary approaches
    Edited by Irene Langran and Tammy Birk

66. **Global Insurrectional Politics**
    Edited by Nevzat Soguk

67. **Time, Globalization and Human Experience**
    Interdisciplinary explorations
    Edited by Paul Huebener, Susie O'Brien, Tony Porter, Liam Stockdale and Yanqiu Rachel Zhou

68. **Localization in Development Aid**
    How global institutions enter local lifeworlds
    Edited by Thorsten Bonacker, Judith von Heusinger and Kerstin Zimmer

69. **The new global politics**
    Global social movements in the twenty-first century
    Edited by Harry E. Vanden, Peter N. Funke and Gary Prevost

Time, Globalization and Human Experience

Interdisciplinary explorations

Edited by
Paul Huebener, Susie O'Brien, Tony Porter, Liam Stockdale and Yanqiu Rachel Zhou

Routledge
Taylor & Francis Group
LONDON AND NEW YORK

First published 2017
by Routledge
2 Park Square, Milton Park, Abingdon, Oxon OX14 4RN

and by Routledge
711 Third Avenue, New York, NY 10017

Routledge is an imprint of the Taylor & Francis Group, an informa business

British Library Cataloguing in Publication Data
A catalogue record for this book is available from the British Library

Library of Congress Cataloging in Publication Data
Names: Huebener, Paul, editor.
Title: Time, globalization, and human experience / editors: Paul Huebener, Susie O'Brien, Tony Porter, Liam Stockdale, Yanqiu Rachel Zhou.
Description: Abingdon, Oxon ; New York, NY : Routledge, 2017. | Series: Rethinking globalizations ; 67 | Includes bibliographical references and index.
Identifiers: LCCN 2016037846| ISBN 9781138697331 (hardback) | ISBN 781315522135 (ebook)
Subjects: LCSH: Globalization–Social aspects. | Globalization–Political aspects. | Globalization–Environmental aspects. | Time–Social aspects. | Time perspective.
Classification: LCC JZ1318 .T56 2017 | DDC 303.48/2–dc23
LC record available at https://lccn.loc.gov/2016037846

ISBN: 978-1-138-69733-1 (hbk)
ISBN: 978-1-31552-213-5 (ebk)

Typeset in Times New Roman
by Taylor & Francis Books

Contents

Contributors

Adam Barrows is an Associate Professor in the English Department and Director of the Institute of Interdisciplinary Studies at Carleton University in Ottawa, Canada. He is the author of the books, *Time, Literature, and Cartography after the Spatial Turn* and *The Cosmic Time of Empire*. He has published widely on time and literature in the journals *James Joyce Quarterly*, *Modern Language Quarterly*, and *Modern Fiction Studies*, among others, and is a recipient of the *Modern Fiction Studies* Margaret Church Memorial Prize.

Brent Ryan Bellamy is a Social Sciences and Humanities Research Council Postdoctoral Fellow at Memorial University of Newfoundland, Canada, and studies the entanglement of literary forms and energy regimes. Bellamy has published work in *The Canadian Review of American Studies*, *Mediations*, *Paradoxa*, *Reviews in Cultural Theory*, and in the essay collection *Green Planets*. He is the editor of two forthcoming projects: a special issue of *Postmodern Culture* on "Resource Aesthetics" (Fall 2016) and an essay collection titled *Materialism and the Critique of Energy* (MCM Prime, Winter 2017). You can follow his academic work at brentryanbellamy.com.

Kevin K. Birth is a Professor of Anthropology at Queens College of the City University of New York, United States. He studies cultural concepts of time in relationship to cognition. His publications and presentations cover a wide-ranging array of topics including chronobiology and globalization, comparative calendars, timekeeping in Roman Britain, culture and memory, cognitive neuroscience, early modern clocks, and ideas about roosters in the Middle Ages. He is the author of four books: *Any Time is Trinidad Time* (University Press of Florida, 1999), *Bacchanalian Sentiments* (Duke University Press, 2008), *Objects of Time* (Palgrave Macmillan, 2012), and *Time Blind: Problems in Perceiving Other Temporalities* (Palgrave Macmillan, 2016).

William D. Coleman is Professor of Political Science at the University of Waterloo and the Balsillie School of International Affairs, Canada. He served three years as the Director of the PhD in Global Governance program and the MA program in Global Governance. He was the Founding Director of the Institute on Globalization and the Human Condition at McMaster University, Hamilton, Ontario, Canada. He had the privilege of being a faculty member in the Department of Political Science at McMaster for 32 years and holds the title of Distinguished University Professor Emeritus at McMaster University. In 2010, he was awarded the title Shakorihonnién:ni (Associate Professor), Deyohaná:ge (Indigenous Knowledge Center), Six Nations Polytechnic, Six Nations Grand River Territory. In 2015, he was awarded the title of University Professor by the University of Waterloo.

Robert Hassan is Associate Professor and Director of the Media and Communications Program at the University of Melbourne, Australia. He has written widely at the intersections between time, politics and information technologies. His most recent book is titled *Philosophy of Media* and will be published by Routledge in 2016.

Wayne Hope is a researcher, lecturer and Associate Professor in the School of Communication Studies at the Auckland University of Technology, New Zealand. His specific areas of research include New Zealand media history and public sphere analysis, the political economy of communication, sport-media relationships, and globalization and time. His research has been published in a range of journals including *Media, Culture and Society*, *International Journal of Communication* and *Time and Society*. He is also co-editor of an online journal entitled *Political Economy of Communication*. Within New Zealand Hope has appeared regularly as a media commentator on television and radio when not writing pieces for *The Daily Blog NZ*. His book *Time, Communication and Global Capitalism* (Palgrave Macmillan, 2016) is now available.

Paul Huebener is an Assistant Professor of English in the Centre for Humanities at Athabasca University, Canada. His book *Timing Canada: The Shifting Politics of Time in Canadian Literary Culture* (McGill-Queen's University Press, 2015) was a finalist for the 2015 Gabrielle Roy Prize.

Susie O'Brien is an Associate Professor in the Department of English and Cultural Studies at McMaster University, Canada. Her publications, on postcolonial ecocriticism, slow and local food movements, resilience and risk, include essays in *Canadian Literature*, *Postcolonial Text*, *Modern Fiction Studies*, *Cultural Critique*, *Interventions*, *Mosaic*, *South Atlantic Quarterly*, *The Review of Education, Pedagogy and Cultural Studies*, as well as (with Imre Szeman) *Popular Culture: A User's Guide*.

Simon Orpana is an SSHRC Postdoctoral Fellow in the Department of English and Film Studies at the University of Alberta, Canada. His writing on subcultures, film and popular culture appears in journals such as *Topia* and *The Review of Education, Pedagogy and Cultural Studies*, and the book *Skateboarding: Subculture, Sites and Shifts* (Routledge, 2015). With Robert Kristofferson, he is co-author of *Showdown: Making Modern Unions* (Between the Lines, 2016), a graphic history of the 1946 Stelco strike in Hamilton, Ontario.

Tony Porter is Professor of Political Science, McMaster University in Hamilton, Canada. His most recent books are *Transnational Financial Associations and the Governance of Global Finance: Assembling Power and Wealth* (RIPE/Routledge, 2013), co-authored with Heather McKeen-Edwards; and *Transnational Financial Regulation after the Financial Crisis* (RIPE/Routledge 2014, edited).

Petra Rethmann is Professor of Anthropology at McMaster University, Canada. She writes and teaches on the relationship between history, memory and politics, as well as on aesthetics, art, political possibility, and imagination. She is the author of *Tundra Passages*, co-editor of *Globality: Frictions and Connections*, and the author of numerous articles that have appeared in edited volumes and in journals such as *American Anthropologist, American Ethnologist*, *Anthropologica*, *Sinij Divan*, *Cultural Critique, Globalizations*, and *Anthropologie et Sociétés.*

Sarah Sharma is Associate Professor and Director of the McLuhan Program in Culture and Technology at the University of Toronto, Canada. She is a faculty member at the Institute for Communication Culture Information and Technology (Mississauga) with a graduate appointment in the Faculty of Information (St George). Her work explores issues related to technology and social difference. She is the author of *In the Meantime: Temporality and Cultural Politics* (Duke University Press, 2014).

Liam Stockdale is a Postdoctoral Fellow at the Paul R. Macpherson Institute for Leadership, Innovation and Excellence in Teaching at McMaster University, Canada. His research interests sit at the intersection of international relations and contemporary political and social theory, with an emphasis on the temporalities of societal governance. He is the author of *Taming an Uncertain Future: Temporality, Sovereignty, and the Politics of Anticipatory Governance* (Rowman & Littlefield, 2016).

Yanqiu Rachel Zhou is an Associate Professor at the Institute on Globalization and the Human Condition and the School of Social Work, McMaster University, Canada. Her major research interests include transnationalism, social policy, temporalities, and global health. She has widely published in edited books and in various peer-reviewed journals, including *Globalizations*, *Journal of Aging Studies*, *Health, Time and Society*, *Social Science and Medicine*, and *Global Social Policy.*

Acknowledgement

This book grew out of several years of discussion about the relationship between time and globalization. This involved three workshops at McMaster University in November 2011, October 2012, and September 2013. The chapters in this volume include papers presented at those workshops, and others that were added subsequently. Throughout these discussions our understanding of time and globalization has been immensely enriched by meaningful and enlightening intellectual exchanges, where participants strove to understand the issues from a variety of perspectives and disciplinary starting points other than their own, as reflected also in Sarah Sharma's commentary that constitutes the final chapter of this book. Dr Sharma's extended closing remarks at the end of our third workshop, drawing on her own *In the Meantime: Temporality and Cultural Politics*, published by Duke University Press in 2014, inspiringly drew together the ideas that had been expressed, linking these to the larger contemporary social and cultural changes to which globalization and time are connected. We are very grateful to all the participants in these discussions and to the anonymous peer reviewers of the book chapters who provided valuable suggestions. We also are grateful to the research assistants who helped move the project along at its various stages, including Diana Gresku, Sean Pittman, Hooria Raza, Sutina Chou and Danielle Blab, and to the Institute on Globalization and the Human Condition (IGHC) and its administrators, Nancy Johnson, Bianca James and Lihua Qian, for their valuable administrative and financial support. Our project was supported by a Social Sciences and Humanities Research Council of Canada Connections grant, an IGHC seed grant, various offices and departments at McMaster University, and Hamilton Artists Inc.

The members of the editorial team at Taylor & Francis, especially Dr Barry Gills, Lydia de Cruz and Nicola Parkin, have been extremely helpful and encouraging in the preparation and publication of both this book and the special issue of the journal *Globalizations*, 13(3) 2016, in which some of the papers presented at our workshops were published. One of those papers, Yanqiu Rachel Zhou & William D. Coleman (2016) "Accelerated Contagion and Response: Understanding the Relationships among Globalization, Time, and Disease," *Globalizations*, 13(3), pp. 285–299, is also included in this volume and is reprinted with permission, as is a paper that was published in a different Taylor & Francis journal, Liam P.D. Stockdale (2013) "Imagined Futures and Exceptional Presents: A Conceptual Critique of 'Pre-emptive Security'," *Global Change, Peace & Security*, 25(2), pp. 141–157.

Throughout the years in which this project has developed, as the initiators of the project and editors of this volume, we have also been grateful for our unusually intense and collegial collaboration, as we grappled with everything from detailed logical issues to challenging and abstract conceptual questions about the intersections of time and globalization. We are delighted that it has resulted in this book, and look forward to ongoing discussion of the issues it raises.

Paul Huebener, Susie O'Brien, Tony Porter,
Liam Stockdale, Yanqiu Rachel Zhou
Athabasca and Hamilton
May 2016

1 Introduction[1]

Paul Huebener, Susie O'Brien, Tony Porter, Liam Stockdale and Yanqiu Rachel Zhou

In *The Condition of Postmodernity* (1992) David Harvey described the contemporary phenomenon of globalization as a process of "time-space compression." In spite of Harvey's title, which gestures towards temporality, much of the scholarship over the next two decades has focused on elaborating the "space" part of the equation. Typically globalization is imagined as bringing together distant places or as the spreading out of dominant practices from the center of the world system to its peripheries. Likewise, popular representations of globalization, both critical and celebratory, have tended to conceptualize it in spatial terms (for example, the global is frequently invoked in opposition to the local, conceived variously as a site of anti-cosmopolitan provincialism, or as the embodiment of authenticity and ecological integrity). Until recently, few critics have addressed globalization as a matter not just of space but also of time. This is the conversation to which this collection is addressed.[2]

Globalization is a contested concept, but has generally referred to the multiple processes (economic, cultural, social, and political) that transcend state borders on a global scale. Regardless of its differential conditions and dynamics in different historical eras (colonial, industrial, and neoliberal), globalization has never consisted of neutral processes, but rather has always been permeated with power, politics and interests. Neoliberal economic globalization since the late 1970s, for example, is not simply about international economic integration, but also about the concentration and accumulation of wealth and maximization of profit for powerful Northern countries and transnational corporations through global capitalist markets (Chomsky, 2000; Rutherford and Davison, 2012). While ongoing globalizing processes have generated a so-called "global society" in which people, places, institutions and nation-states have become increasingly interdependent, it is always important for us to pay closer attention to the wide variety of globalization experiences at all levels, including the uneven distribution of its benefits and costs. It is also crucial to analyze critically the significance of "globalization" as a model of contemporary economic, political, and social organization. Globalization can be read as a phenomenon or set of processes, but it can also be read as an idea, "a concern and desire," as Petra Rethmann puts it in this volume, that helps to illuminate certain aspects of contemporary life, and conceals others.

Turning to the question of time, Rita Felski has persuasively noted that time "is a concept of enormous complexity, including questions of measurement, rhythm, synchronization, sequence, tempo and intensity. It spans the personal and the public, work and leisure, the instantaneous and the eternal, intimate relations and global structures, everyday life and conditions of extremity. It exists at many different levels and is experienced in radically divergent ways" (Felski, 2000, pp. 16–17). Accordingly, it is our position that proposing any specific articulation of what we consider time to be might prove counterproductive to the project at hand. Indeed, not only have two millennia of philosophic inquiry struggled to elucidate a clear ontology and phenomenology of time,[3] but attempting to do so here would work against the critical ethos that underpins this project, since it would serve to foreclose one of the key aspects of the terrain we are interested in exploring. In other words, the contextual variability of humanity's understanding and experience of time is one of the most interesting aspects of the contemporary intersection of temporality and globalization. Rather than addressing the question of temporal ontology directly and specifically, therefore, we broadly concur with Thomas Allen, who argues that "time is defined in relation to the purposes it serves. Hence, we can never say what time is. Instead, we can only ask what kinds of worlds different forms of time make possible, and what interests are served by the creation of such worlds" (Allen, 2008, p. 217). These are precisely the sorts of questions that the contributions to this volume consider, and collectively they develop a rich exploration of the complex interrelationship between time and globalization in the contemporary context.

Within this open approach to time we can note, however, one assumption about it that is consistent with the approaches represented in this book. That is that while time may interact with certain features of the material world, such as the rotation of our planet or the aging of our bodies, it is also always socially constructed. A second assumption that informs many of the chapters in this collection is that we can distinguish, at least to an extent, between time and temporality. Time is the more general phenomenon, which can include, for instance, its measurement in devices such as clocks (time reckoning), or our understanding of its role in creation of the universe, far from our daily experience. In contrast, temporality is the lived experience of time (Hope, this volume). This can be, for instance, a visceral awareness of rhythms (Barrows, this volume) such as heartbeats, which operates in complex relation with other modes of temporality, such as social dynamics of hope (Rethmann, this volume) or anticipation (Stockdale, this volume). Each of these temporalities may or may not rely upon or match up with aspects of time that are less directly connected with that lived experience.

One reason to study the interaction of globalization and time is the multiple examples of their interrelationship historically. Time and globalization each have their own complex social histories, yet the two domains have always been intertwined. Even Eratosthenes's calculation of the circumference of the globe in the third century BCE required the coordination of simultaneous measurements of the sun's rays in different locations (Roller, 2010, p. 13). The empires of antiquity, meanwhile, had to coordinate between various different local timekeeping systems.[4] A more modern relationship between time and broad geographical connectedness becomes visible in Europe in the 18th century, when the spread of mechanical clocks and print culture fostered a shared social feeling of simultaneity and created what Benedict Anderson calls an "imagined community," a conceptual nation in which individuals move forward together through time (Anderson, 1991, pp. 24–25). Timekeeping technologies also played a key role in European intercontinental exploration and colonization (see the Materiality section below); for seafarers, taking accurate longitudinal measurements in the open ocean required knowing the vessel's east–west spatial distance, and thus daylight time differential, from a known observatory such as Greenwich, a feat that was nearly impossible to accomplish until the English horologist John Harrison completed a chronometer in 1759 which could maintain its accuracy during rough intercontinental voyages (Sobel, 1995, p. 106). Thus, precise timekeeping across global distances was spurred by, and in turn facilitated, the globalizing processes of exploration, cartography, resource exploitation, and colonization. What is more, globalization and colonization increasingly make visible the cultural and ideological aspects of time, as the imposition of Western forms of time reckoning – clock time and weekly working days as well as concepts such as progress, punctuality, and productivity – onto European colonies suggests that imperialism is as much a temporal process as it is a spatial one (Nanni, 2013). Peter Galison notes that the massive project of laying intercontinental telegraph lines across the Atlantic Ocean in the mid-19th century created more accurate clock synchronizations between Europe and North America (Galison, 2003, p. 104), so that, in Christopher Dewdney's words, "the concept of a global 'now' inserted itself into the average citizen's consciousness, at least in Western nations" (Dewdney, 2008, p. 64), a development quickly followed by the invention of the space-time-collapsing telephone. The familiar global network of standardized time zones was established in the late 19th and early 20th centuries, based partially on a proposal by the Canadian Sandford Fleming, who, frustrated by the use of uncoordinated timekeeping systems in countless different localities, had identified an urgent need for what he called "a comprehensive view of the globe in considering the question of time-reckoning" (quoted in Blaise, 2001, p. 35). Adam Barrows notes, however, that the ability of global standard time to enable "the precise coordination of global activity" goes hand in hand with "the demands of international commerce and military hegemony" (Barrows, 2011, p. 4).

Thus, while discussions of time and globalization early in the 21st century may emphasize the sense of a shared global present brought about by electronic networking, and may highlight the disjunctures and power relations at stake between this single global temporality and multiple local temporalities, complex and contested visions of globalized temporalities are by no means an invention of the computer age. The contemporary links between time and globalization represent only the latest developments in a long process.

Turning from history to more contemporary analysis, there is a rich and extensive literature on time that has often not considered time's relationship to globalization, but nevertheless suggests multiple connections between time and globalization. This literature suggests that time plays a central role in organizing human experience, and thus intuitively time must be entwined with globalization. For instance, the sociological literature on time has identified "social acceleration" as a key feature of modernity. Rosa (2013, p. 305) notes that contemporary social acceleration involves destabilizing intra-generational change:

> this leads to an altered form of time experience and hence to new forms of personal identity and political self-understanding that have come to be defined by the metaphors of *flows* and *fluids*. As a result of the erosion of the institutions of the welfare state, the experience of a progressive development along pre-established tracks is gradually transitioning into a perception of undefined situations with high, unforeseeable rates of change.

Similar effects to these have been associated with globalization. From Rosa's perspective globalization, and the greater speed with which people and things travel and connect across the globe today, are part of these larger changes in time (Rosa, 2013, pp. 211–223, 300).

Much as with globalization, this social acceleration can threaten slower traditional mechanisms for democratic expression, benefiting more powerful mobile actors at the expense of others. For instance, Sassen (2000, p. 224) has analyzed the interaction of slower national and faster global temporalities, involving "a shift from a bureaucratic temporal order, in principle subject to the slow moving and rule-bound public accountability of governmental processes, to the accelerated dynamic of private 'regulatory' functions and markets." This project emphasizes that a critical understanding of the dynamic relationship between time and globalization would not only facilitate innovative thinking about globalization; it would also foster our imaginations about the alternatives that may lead to more socially just and sustainable futures.

Indeed, actively managing the future – whether to bring about idealized visions or to preclude the irruption of potential catastrophes – is a powerful aspect of contemporary human experience (Giddens, 1999). As discussed further below, this has found expression in the proliferation of anticipatory modes of action across a wide array of issues that are also global in scope – from environmental governance to the conduct of warfare. In the latter case, for instance, the accelerative dynamics of globalization are particularly acute; advances in weapons technology, such as unmanned drones, combined with the post-9/11 imperative to stop the proverbial "next attack" before it occurs, create a global security regime in which powerful states can deploy military force nearly instantaneously across the globe, with a view to pre-empting potential threats before they emerge (Stockdale, 2016). In this sense, understanding contemporary societal responses to the seemingly omnipresent specter of transnational terrorism also requires reckoning critically with the intersection of time and globalization.

In light of such considerations, Sharma (2014, p. 142) has compellingly argued for recognizing a pluralistic "temporal public," moving beyond thinking of publics as purely spatial: "we need a radical politics of time and space that hinges upon temporalizing the spatial categories of democracy. … This would be a politics for the meantime – keenly aware of the hold that structures of power have upon time." Bringing in temporality helps make power, interdependence, movement, and contingency more visible in spaces that might not otherwise be recognized as public. This then has implications for how we can reconstruct these: "a temporal perspective does not try to create more free time; it strives to free time from this fixation. It reimagines time, not as being singularly yours or mine for the taking, but as uncompromisingly tethered and collective" (Sharma, 2014, p. 150). This can be local, and most definitely global as well.

The distinctive focus of this volume, then, is the interactions between globalization and time and their relationship to humans and the world we live in. Despite their frequent mischaracterization as purely natural phenomena, both globalization and time are densely entangled with human imagination and sociality. Humans intervene in both, to varying degrees, and these interventions in turn shape the ideas and practices of humans. As in all human affairs, these interventions involve the production and deployment of power, but because the interactions between globalization and time are not always recognized, their distinctive connections to power can be obscured as well.

The interplay between these varying aspects of the human experience of time and globalization is a reason to understand this through the type of interdisciplinary approach that this book takes. Rosa (2013, pp. 282–286) has noted the effects of social acceleration in social structures, culture, personal identity, and technological and economic interactions with nature. Globalization, too, has this multidimensional character. The chapters that follow bring together approaches from diverse humanities and social science disciplines, analyzing the human experience of time and globalization in literary fiction, cinema, economic and technical change, the spread of disease and the policy response to this, the conduct of war and the pursuit of (trans)national security, and the constraints and opportunities for political action and imagination more broadly.

As abstract epochal forces, globalization and time display similar properties that can be fruitfully compared. They are inseparable in the sense that space and duration are both needed for things to exist physically, and in the way that distance can be measured in the time it takes to get there. However, they each have distinctive properties, and it is important to identify these and explore their interactions. In the remainder of this opening chapter we focus on four key areas of interaction: chronopolitics and temporal inequalities; heterotemporalities and resistance; materiality; and global futures.

Chronopolitics and temporal inequalities

In his seminal work *Time and the Other*, Fabian (1983) argues that the politics of imperialism and other power relations are bound up with cultural experiences and uses of time, and that "*geopolitics* has its ideological foundations in *chronopolitics*" (p. 144, original emphasis). Global colonization often depended upon, and reinforced, what Fabian calls the denial of coevalness – that is, the denial that all existing people or cultures are contemporaneous with one another – a process through which "one assigns to the conquered populations a *different* Time" (Fabian, 1983, p. 30, original emphasis). Thirty years after his writing, the study of chronopolitics – of the power relations inherent in contested notions of time – remains central to our understanding the processes, evolution and consequences of globalization.

Against the background of industrialization, for instance, the introduction of "universal" clock time has not only marginalized other temporal logics associated with other cultures, places, histories and human experiences, but it has also consolidated the hierarchical relationship between Western "modern" and non-Western "pre-modern" times (Hassan, 2005; Rosa, 2009). The denial of coevalness, like other forms of temporal domination, occurs not only on a global scale, where colonizing forces used the "primitive" nature of indigenous populations as a justification for dominating and "civilizing" entire cultures; it also happens *within* societies, often in ways that correspond with discrepancies in socioeconomic status (Fabian, 1983). Along this line, impoverished people are often said to have "fallen behind," a temporal distancing metaphor that implies a failure on the part of the individual to "keep up" with society's progress. This can also be taken as an example of how our thinking of localized issues is mediated by the politics of time.

As well, the politics of speed is especially important for considerations of chronopolitics in the context of contemporary globalization, when society witnessed "a significant acceleration in the pace of life concomitant with a dissolution or collapse of traditional spatial co-ordinates" (May and Thrift, 2001, p. 7). Contemporary life is increasingly defined by speed, by considering the ways in which high-speed technologies modify relationships and power structures on the global scale, and investigating how high-speed actors are increasingly able to dominate slower ones (Rosa and Scheuerman, 2009; Virilio, 2000). Being prevalently used in warfare, media, and other cultural realms, technologies that operate through high-speed networks create "the possibility to act instantaneously, from afar" (Virilio, 2000, para. 12, lines 4 5). We can see this, for example, in the US military's use of unmanned drones flown by operators in the United States to conduct real-time airstrikes on the other side of the world (Stockdale, this volume), or in the "high-frequency trading" practices used by powerful financial actors that operate around the clock to translate the speed advantages of fiber optic networks into significant gains across global financial markets (Porter and Stockdale, 2015). Because "everything, [including globalization,] is about speed and real time," Virilio

(2000) argues, "there is a real need not simply for a political economy of wealth but also for a political economy of speed" (para. 12, lines 19–20). Linking people's access to speed with their access to power, Bauman similarly notes that people "who come closest to the momentariness of movement are now the people who rule. And it is the people who cannot move as quickly, and more conspicuously yet the people who cannot voluntarily leave the place at all, who are ruled" (Bauman, 2000, pp. 178–179). That is, differential access to instantaneity and rapid adaptability may also suggest the differentiation between domination and subjugation. For instance, the hegemony of an accelerative logic that values efficiency and productivity has prioritized economic time over other forms of time such as cultural time, family time and leisure time, which are integral to social and human well-being and economic sustainability. Hochschild (2005, p. 339) even reports that family life itself in modern America is "under strong pressure to incorporate aspects of market culture" in order to bring about greater "efficiencies" in personal relationships.

The politics of contested notions of time and speed are significant in human relationships with the natural world. Because contemporary capitalist societies privilege short-term profitability and restrictively linear modes of thought, these temporalities form "a central part of the deep structure of environmental damage wrought by the industrial way of life" (Adam, 1998, p. 9). The influence that the politics of speed and linear productivity can have on environmental sustainability is visible, for instance, in the adoption of what Canada's former Finance Minister Jim Flaherty calls an "accelerated regulatory system" (Paris, 2012). Touting the necessity of such a system for the economic benefits of global energy shipments, Flaherty announced in 2012 that the Canadian environmental reviews process for major resource development projects, such as the controversial Northern Gateway oil pipeline, would be limited to a maximum period of two years, replacing the previous limit of six years (Paris, 2012). The imposition of such frames of time privileges the short-term profits of global corporations at the expense of the long-term health and security of the human and non-human agents that bear the consequences of rapid development. It also fails to account for the multiple, complex, non-linear timings and temporalities that exist at many different biological and ecological levels.

As much of the above discussion suggests, the matter of temporal inequalities in the context of economic globalization is also closely connected to chronopolitics. In her article "The Global City," Sassen (2000) argues that the intersection of two coexisting temporalities – an old or collapsing temporality (historical time) and a new temporality (the time of economic globalization) – leads to the partly de-nationalized temporalities that drive and constitute economic globalization. Hope (2009) also points out that the structures and activities of global capitalism exacerbate temporal disjunctures between the temporalities of the nation-state, the economy, and society. For many Southern countries where global capitalism and neoliberal agendas have been imposed, hierarchies of temporal power are not only a reflection of the international socioeconomic inequalities that are historically constructed, but have also become a foundation on which further inequalities – in such forms as differential economic development, political stability, social welfare, and environmental sustainability – are reproduced and sustained. In the context of emerging infectious diseases (EIDs) such as SARS, avian flu, H1N1, and Ebola, however, the temporal inequalities across affected countries – such as their differential speed to respond to a disease – may lead to a global health crisis that would directly challenge global capitalism itself (Zhou and Coleman, this volume).

In addition to the macro-structural level at which hierarchies of temporal order often reflect asymmetries of power, another layer of temporal inequalities concerns the complex impacts of dominant temporal orders on individual people's lives. The broad social politics of speed and efficiency have significant material impacts on individuals, since members of particular segments of society do not have the same capacity (or desire) to respond to the associated demands due to cultural particularities, geographical restrictions, and/or unequal access to resources, including information and communication technologies (ICTs) (Rosa and Scheuerman, 2009). While global capitalism has notoriously exacerbated socioeconomic inequalities among people and countries through trends such as income and wealth polarization in both the global South and the global North, time itself as an economic resource also functions as an axis of social inequalities, in complex tension with other axes such as gender, class, race, ethnicity, age, ability, sexuality, and access to social capital. When women's participation in the labor force is increasingly promoted by Western welfare states, for instance, "time deficit" and work–family conflicts then become part of these women's daily struggles. The conflicts could be "resolved" only if their care work can be downloaded to their family members (usually grandmothers) or outsourced to paid care workers (primarily women from less wealthy countries) through the global care market, which has generated further inequalities based on gender, age, and geography (Zhou, 2015). When the wages for unskilled workers are driven down by "international competition," similarly, one of the "strategies" for these workers to make ends meet involves squeezing their family and leisure time in order to take on multiple jobs, or jobs requiring "temporal flexibilities" (Hassan and Purser, 2007, p. 73).

Investigations of chronopolitics and temporal inequalities reveal that the question of who has the power to legitimatize, impose, and manipulate certain temporal frames and ideologies is a vital one for considerations of globalization. Assessing the role of chronopolitics in transnational settings, for example, Porter and Stockdale (2015) argue that temporalities are strategically manipulated at a macro level "in the accumulation of power and wealth or in the resistance to such accumulation" powerful actors are able to "harness or manipulate particular temporalities for strategic purposes" through the creation of dominant "temporal systems" that "condition how time is apprehended and experienced in human affairs" (p. 4). Internalizing the dominant temporal framework that advocates economic acceleration, at a micro level some individuals and families have used migration as a strategy to manage or overcome the "temporal gaps" between places (Rosa and Scheuerman, 2009; Zhou, 2015). Nevertheless, the global flow of people has also brought about the coexistence of multiple temporal-spatial relationships that lay a foundation for counteracting the hegemonic temporal logic and its embedded power structure.

To gain a critical understanding of temporal inequalities and their connections with globalization, therefore, we need to go beyond the Euro-American-centric and economic-centric perspectives, and listen to the voices of the marginalized and the "other," considering such issues as the temporal struggles of the poor (Coelho, 2003), the time experiences of those with chronic diseases (Cazdyn, 2012), and "queer time" (McCallum and Tuhkanen, 2011). Such analyses can work to identify the relationships between the micro and the macro, and to evaluate the potential for transformation and resistance. Several of the articles collected here take up the matter of strategically manipulated temporalities by examining the concepts of chronopolitics and temporal inequalities within contemporary economic, political, cultural, social, and ecological configurations. They collectively show how revised evaluations of the politics of time are critical for understanding both broad theoretical discussions and specific practical approaches to globalization.

Heterotemporalities and resistance

Any homogeneous form of temporality that is imposed onto disparate situations and populations tends to silence, or reduce the possibilities for imagining, forms of political sociality and temporality other than the normative one that seems to be prescribed by corporate, professionalized, technocratic, and democratic disciplinary systems. As some of the factors mentioned above make clear, though, while the contested politics of global, linear, or profit-based time often facilitate the entrenchment of power, the manipulation of those who are deemed "slow" or "primitive," and the denial of long-term ecological and generational responsibilities, these contested temporal politics also create possibilities for various forms of resistance, disruption and contradiction.

Given the global prevalence of the cultural ties between time and money, the matter of wage labor presents a particularly potent site for both temporal domination and temporal resistance. As Jeremy Rifkin observes, "in industrial cultures, the poor are temporally poor as well as materially poor" (Rifkin, 1989, p. 92); low wages and extremely long working hours tend to go hand in hand. In the late 19th and early 20th centuries, workers involved in the labor movement demanded shorter working hours; one popular slogan in the United States was "[e]ight hours for work, eight hours for sleep, eight hours for what we will" (Levine, 1997, p. 141). This trend spread throughout many industrialized countries with the help of organizations such as the

International Workingmen's Association. At the turn of the 21st century, with organized labor in severe decline, the temporality of work remains a subject of contestation. Calls to adopt a four-day work week persist alongside more radical moves such as Timebanking – a loosely structured organization whose members barter their own labor in equally valued one-hour segments. At the same time, however, the apparently irresistible pull of continuous network connectivity through smartphones and other devices has, for many workers, increased time pressure by leaving no portion of each 24-hour period entirely free from work duties. As Hartmut Rosa (2013, p. 192) argues, traditional forms of time management and planning in many people's lives are "being replaced by flexible time arrangements that (similar to the processing of digital computers) involve microtemporal oscillation between the demands of distinct functional spheres that are all running as 'non-stop' enterprises." Thus, 21st-century conflicts between capital and labor remain closely tied to experiences of time, yet require increasing attention to forms of connectivity and apparent simultaneity.

Non-stop connectedness, along with other aspects of global time, operates around, without ever totally subsuming, what Bliss Cua Lim (2009) calls "immiscible" times. Such temporalities, which Victor Li (2015) suggests include the temporalities of human bodies and other "objects and lifeworlds that do not belong to the futural thrust of cyber-capitalism" (p. 6), exert substantial influence but are not well accounted for in studies of globalization that emphasize universal (if uneven and discontinuous) processes of development, acceleration and synchronization. Ernst Bloch's (1991) concept of "synchronous nonsynchronicity" goes some way toward acknowledging the co-presence of multiple temporalities, which may take the form of the persistence of remnants of the past haunting the present. Such haunting can take interesting and strikingly literal forms, such as bouts of spirit possession experienced by Malaysian women factory workers in Japan (Ong, 1987), or recurrent encounters by migrant subway construction workers in Bangalore with ghosts from villages they left behind (Mittal and Raghunandan, 2012). While susceptible to different readings, these experiences arguably support Rita Felski's wariness of embracing fully the evolutionism that informs Bloch's (Marxist) project, reducing temporal differences to stages in a universal narrative of progress (Felski, 2000, p. 22).

Not only does the vision of universal progress risk reinforcing the colonialist paradigm of the denial of coevalness outlined by Fabian, but it also fails to take account of the "multi-dimensional" temporalities of "nature, the environment and sustainability" (Adam, 1998, p. 9). These temporalities are irreducible to culture, while remaining entangled with it in complex, sometimes contradictory ways. Think, for example, of the "leap second" that we need to add, every now and then, to our official time measurements and computer software in order to keep our culturally constructed clocks in synch with the wobbly Earth. Or consider the ways in which the global dependence on petroleum for fueling everyday economic activities masks the slow geological temporalities of fossil fuel creation and the long-term ecological consequences of climate change. On the one hand, the temporalities of the non-human world – the seasons, diurnal rhythms, and the lifespans of living things – are mediated and altered by human technology; indeed, the disruption or intensification of these processes is a hallmark of modernity, reaching its apex in the (unintentional) anthropogenic modification of the climate (Adam, 1998; Chakrabarty, 2009). On the other hand, the fallout of nuclear and other environmental accidents expresses a complex, non-linear temporality that highlights the incapacity of human culture fully to apprehend, let alone control, the temporal currents of the physical world (on responses to that incapacity, see the Global futures section, below).

It is easy to see the relationship between human and non-human temporalities as an increasingly fraught and combative one, with perilous consequences for all living things. While humans do find resonance with non-human times through everyday practices such as gardening (Adam, 1998, p. 95), some activists and critics call for more radical temporal recalibrations. For example, John Urry advocates for the adoption of what he calls "glacial time," a consciously slow form of social engagement which "resists instantaneous time and seeks to slow time down to 'nature's speed'" (Urry, 2009, p. 194). While this project does not account for the fact that "nature" contains many differently paced temporalities, its critical resistance to intensively speed-based social politics is compelling, and is reminiscent of other aspects of the Slow Movement, most notably Slow Food, whose advocates seek to allow culinary production and pleasure to proceed at their own relaxed pace in defiance of the culture of ubiquitous fast food (Parkins and Craig, 2006). Such practices can demand a degree of leisure that restricts, on the basis of class, the ease of their adoption (see the Chronopolitics and temporal inequalities section, above). Even those communities whose traditional cultures follow more closely the rhythms of the natural world, including urban migrants from rural areas and aboriginal people living both off and on reserve, sometimes face significant obstacles to opting out of the "fast" lifestyle represented by global time. In spite of these challenges, traditional indigenous cultures often continue to insist on the interdependence of human and more-than-human temporalities, and to retain a corresponding recognition of multiple scales of time, from the cosmological to the ecological, and the interdependence of time and place (see for example, Sioui, 2008).

These distinct features of non-Western cultures are important to acknowledge. However, many critics (Gupta, 1992; Felski, 2000) caution against reifying a split between Western time, often described as linear, empty, homogeneous, and progressive, and non-Western (cyclical, heterogeneous, "natural") time. For example, Akhil Gupta describes Indian villages:

> where work is primarily agricultural, where many men wear watches, where radio programs, buses, and trains work on schedules made up in abstract time (or at least are supposed to), where reincarnation and

> transmigration occurring across incredibly large temporal units are considered a commonplace feature of everyday life.
>
> (Gupta, 1992, p. 202)

Incommensurable temporalities exist within as well as between so-called developed and developing societies. Barbara Adam argues that critical analysis of our own cultural assumptions must form the basis for resisting oversimplified visions of social time; only by highlighting "the immense variability and complexity of social time" can we "expose the poverty and inadequacy of models that force us to choose between time and temporality, clocks and natural rhythms, linearity and cyclicality" (Adam, 1994, p. 508).

A key objective of our project is to articulate and to affirm temporal variety, what Lefebvre (2004) calls the diverse "polyrhythmia" of everyday life, as a counter to hegemonic global time. However, this is not simply a matter of defending the individual and the aleatory against a universal tide of history. Heterotemporalities are not automatically aligned with resistance any more than global convergence equates with oppression. As Manuel Castells has noted, the new social movements that have emerged to counter the accelerated connections and functional logic of the network society include religious fundamentalism and nationalism as well as environmentalism and feminism (Castells, 1997). Moreover, notwithstanding our caveats against simplistic social evolutionism, we cannot dispense with the pragmatic value, if not the measurable validity, of the concept of progress. As Meaghan Morris has noted, "rhetorics of evolution, elevation, advancement, 'upward mobility,' and cohesively linear time have served with as much tenacity to articulate the aspirations of radicals, reformers and dissenters as they have to validate and channel colonial power" (Morris, 1998, p. 11). This kind of cohesive, forward-moving and universal mobility is arguably one of the temporal models that are threatened by the "instantaneous time" of globalization. Indeed, as Eric Cazdyn and Imre Szeman (2012) have noted, a key feature of the present moment, and one of the principal impediments to achieving progressive change, is the belief that the temporality of globalization, whose contours are defined by capitalism, is natural and inevitable: a time *after* globalization has become almost unthinkable.

This cognitive impasse highlights the key role of imagination in the production and contestation of temporalities. Narrative is an important site of resistance against dominant temporal modes. Paul Ricoeur has shown that "time becomes human time to the extent that it is organized after the manner of a narrative" (Ricoeur, 1984, p. 3) – in other words, temporality for human beings is inevitably articulated and experienced through various forms of narrative (Ricoeur, 1984, p. 3). If global time is essentially a story that is told with the aid of electronic networking and global transportation and financial networks, opportunities exist for reworking this narrative to reflect alternative priorities. Postmodern fiction in particular has been adept at questioning modern temporal and narrative assumptions such as the idea that time moves forward in a linear progression (one in which, colonialist ideology has it, some cultures are more advanced than others). The questions of whose temporality, whose articulation of the past and the future, or whose temporal and financial metaphors gain purchase and become reified, are tied inevitably to the question of whose narratives become most prevalent on the global stage.

Materiality

Our experience of time involves a relationship between human consciousness of time and material rhythms (Barrows, this volume). Some of these material rhythms, such as the Earth's rotation, are independent of humans. Others, like a heart beat or a life cycle are associated with the human body. Yet others, such as the rotation of a clock's hands, are objects that are constructed by humans, constituting what might be termed temporal artifacts (Birth, this volume). The interactions between human consciousness and material rhythms have always been central to social organization. Historically these interactions have become increasingly complex, and increasingly detached from the natural world. Electric light, irregular work schedules, and sleep medications break our physical synchronization with the rhythmic passage of day and night (Crary, 2013). More generally, the interplay between human bodies, the physical properties of our planet and the non-human life forms on it is closely linked to time and globalization, as with the temporal aspects of global climate change, the speed of global pandemics (Zhou and Coleman, this volume), or the potential and limits of the ability of technical artifacts to allow some humans to escape the slowness associated with the materiality of the human body and its built and natural environments.

The interactions of time and globalization have always been linked to technical artifacts, as with the measuring devices that created global time zones in the 19th century, or today's ICTs that seem to shrink the world and accelerate daily life (Hassan and Purser, 2007). This complexity has been enabled by technical advances in time reckoning, such as the invention of clocks and schedules, through the creation of other infrastructures, such as faster roads, that produce or shape temporalities (Thrift, 2004). Advances in time reckoning enable more social complexity. Complex processes require coordinated sequencing, facilitated by temporal artifacts. This connection between temporality, social complexity, and temporal artifacts is tied to globalization as well. The temporal coordination enabled by temporal artifacts makes the intensifying of connections across borders possible, as exemplified by airline or ocean shipping schedules.

There are important debates about the effects on human freedom of the temporal artifacts that produce or manage time. One question is whether these artifacts are tools that humans control, or instead have become autonomous systems that control humans and restrict our freedom (Stiegler, 1998). Another question is whether this restriction of freedom is a property of temporal artifacts themselves, or if instead temporal artifacts are used by dominant actors to control and exploit subordinate actors (Porter and Stockdale, 2015). These debates are similar to debates about money, another ubiquitous abstract system. As these temporal artifacts have become transnational, these debates resonate with debates about globalization. Is globalization about new opportunities provided by social complexity, or does it erase local freedoms? Is it an expression of human creativity, or a mechanism for powerful actors to reinforce their dominance? Temporal artifacts are an often unrecognized infrastructure of globalization, and thus it is not surprising that they provoke similar ambivalence.

It is obvious that in recent years temporal artifacts have experienced the types of epochal transformations that have been evident in digital technologies more generally, and this has very significant implications for their interaction with globalization. These changes correspond to the even broader transformations that have been identified with the shift from Fordism to post-Fordism. Fordism – the social relations associated with factory-based mass production – works through stable hierarchies within national territories, while post-Fordism is associated with precarious, flexible, transnational networks. The shift from the stopwatch, which was emblematic of the hierarchical control of assembly line workers' bodies, to the robot, with more flexible, decentralized and autonomous timekeeping capacities, exemplifies the transformation of temporal artifacts associated with this change. A similar transformation is visible in the interaction of the time consciousness of individuals and the temporal effects of the shift from broadcast media and cinema to new social media. Hansen (2009) has argued that the controlling, homogenizing and universalizing effects of this new media's temporalities identified by Bernard Stiegler, which displace human imagining in pre-set temporal blocks and structures, are giving way to a far more open temporality, which specifies a before and after in only a very minimal way, enabling varied locally constructed temporalities. While this may be too optimistic, it certainly points to the importance of considering the role of digitized temporal artifacts in debates about the effects of globalization.

A key concern associated with digitized temporal artifacts is whether they hobble our individual and collective capacity to manage our fate by erasing past and future and reducing the domain of action to an instantaneous present (Leccardi, 2007; Lübbe, 2009; Scheuerman, 2009). Certainly digital technologies, such as risk modeling software, have also been implicated in a shift from the past and tradition to the future as the location in which to anchor our destinies, as discussed in the section on Global futures, below. However, the sheer volume and speed of the flow of images, data, and ideas that we experience can make the dominance of the present overwhelming. Often environmental and other long-range negative effects of current practices are obscured (LeBel, 2016). Activists seeking to challenge dominant actors can be attached to the immediacy of spontaneous action, forgoing the more systematic initiatives that can be associated with strategic planning or visualizations of common futures for fear of their oppressive effects (Petrick,

2015; Conway and Singh, 2009), although there is a countermovement that advocates slowness (see the Heterotemporalities and resistance section, above). However, as Sharma (2014, p. 134) has argued, advocating slowness can obscure the privilege it can entail, and the constraints that the materiality of everyday life imposes on most people: "many of these material relations are fast paced, mindless and mundane; slowness hides the rhythm of material relations, the very condition of its possibility."

At a global scale, problems such as climate change become more dauntingly complex, and the loss of the capacity for collective action that social acceleration can involve is problematic, although technological advances in measurement and analysis can also make such changes more visible. Similarly, digital media can give distant events the immediacy that provokes action. Again, the effects of digitized temporal artifacts are ambiguous, but crucial to consider nonetheless.

Although digital technologies promote instantaneity, artifacts in general also play an important contrary role in creating continuity. As Latour (1990) has put it, "technology is society made durable." The preservation of certain cultural interpretations in museums or archives, or the materialization of law in written constitutions or legislative buildings reinforces their durability and power. Today our interactions are increasingly mediated through objects of various types, challenging the way that the "dominant conception of identity as 'presence' is manifested in terms of the materiality of the here and now" (McQuire, 1999, p. 154). This is especially evident in global interactions, although it is important also not to underestimate the entanglement of digital temporal artifacts with local neighborhoods (Crang et al., 2007). Technical objects become even more important in determining which practices, memories or ideas will be sustained over time and which will fade.

The interaction of power and technical artifacts with regard to spatiality in globalization has been recognized with regard to the way that these artifacts allow the projection of power over distance, from the role of navigation equipment when Portuguese sailing ships first traveled to Asia (Law, 1986), to the US-based piloting of drones operating in the greater Middle East today (Stockdale, this volume). The importance of speed in war and finance has also been recognized (der Derian, 1990; Lewis, 2014). Space and time are inseparable dimensions of speed. Powerful actors enhance their power globally by controlling temporal systems, such as schedules, and by creating faster institutions, production processes, and communications systems (Porter and Stockdale, 2015). Sharma (2014, p. 139) notes that "temporal architectures consist of technologies, commodities, policies, plans and programs as well as the labor of others. Whether manifest in napping pods, yoga at the desk, or a slow-resort holiday, this temporal architecture continuously confirms and maintains the same structures of power that drain, tire, and exploit other people's time, while elevating one group's sense of temporal importance." These architectures, as with exclusive resorts or fast lanes in airports, can have a global character: "hotel maids receive training to deal especially with the jet-lagged … call-center employees straddle multiple time zones in order to uphold the work and leisure demands of people on the other side of the globe" (Sharma, 2014, p. 139). Overall, then, temporal artifacts, technologies, and materiality more generally play a central role in the interactions between time and globalization.

Global futures

One of the more interesting aspects of the contemporary intersection between time and globalization is the paradoxical duality of our relation to the future that has emerged in this context. A number of prominent theorists have observed that contemporary society is becoming more explicitly future-oriented, where we seek to control risks but find this increasingly challenging (Giddens, 1999). Ulrich Beck has famously described this phenomenon in terms of a "world risk society," wherein social and political processes are increasingly concerned with managing the "manufactured uncertainties" that have accompanied the onset of post-industrial late modernity (Beck, 1999). Thinkers like Beck point to the emergence of novel hazards – such as environmental degradation, global financial instability, and transnational terrorism – whose management is rendered particularly difficult by their operation beyond traditional mechanisms through which the social world is rendered intelligible, calculable, and governable. The global or "systemic" scope of such risks thus seriously inhibits our capacity to bring the irruptive contingencies of our world under control (Beck, 2002, p. 41). Some also see this phenomenon as illustrating a broader trend, whereby the Enlightenment dream of human mastery over the natural world is becoming increasingly illusory, and where temporal rather than spatial concerns are coming to dominate the contemporary human experience. Such a view is captured well by social theorist Armin Nassehi, who argues that "modernity promised the capacity to shape and control the world and time … [b]ut in late modernity, time itself has come to destroy the potential for any form of social or substantial control" (quoted in Rosa, 2009, pp. 102–103). From this perspective, in other words, a globalized world is one where the radical uncertainty and catastrophic potentiality of an ungovernable future looms large over the lived experiences of an ever more precarious present.

Yet if we have entered an era characterized by heightened awareness of an increasingly unstable future in general, this is in many ways curiously paralleled by a broad-ranging blindness to the future in terms of specific patterns of behavior and action. In this respect, the globalized world from which the aforementioned "systemic risks" have emerged was in large part created by the proliferation of a socioeconomic order that prioritizes growth over sustainability and short-term profit over long-term investment. The result is that even if such hazards as climate change and financial instability are recognized as portents of catastrophe, the ethos and incentives of the neoliberal present simultaneously militate against meaningful action to address serious risks located in a far-off future. Barbara Adam makes this point well by emphasizing that the Western "liberal democracies' unequivocal commitment to a global economy and the quest for sustained economic growth" means that "future people and environmental sustainability ... are not readily encompassed within the boundaries of their concerns" (Adam, 1998, p. 15). The persistent difficulty of enacting consequential climate change legislation in both national and global political contexts is perhaps the most overt illustration of this point, as governments facing short-term election cycles are hesitant to act in the face of entrenched socioeconomic interests arrayed against absorbing any present costs in the name of future sustainability. Moreover, as discussed in the preceding section on Materiality, the technologization of all aspects of life has exacerbated these broader societal tendencies at the individual level, as dynamics of social acceleration and informational overload render it increasingly difficult to think outside the immediacy of the now and seriously reflect upon the future consequences of present action.

It should be noted, however, that apparent gulf between orientations toward the future and attendant behaviour may be in the process of being bridged. Indeed, contemporary trends in both societal governance and cultural production have begun to evince a concerted desire to recapture some semblance of control over the unfolding of the future. This is visible in the rise of discourses of "risk management" across myriad areas of human affairs (Power, 2004), and the associated emergence of anticipatory logics of action aimed at actively pre-empting specific contingencies in a variety of governmental contexts (Anderson, 2010; Stockdale, 2016). The contemporary popularity, in both the public and private sectors, of so-called "scenario planning" – historically the arcane preserve of Cold War national defense intellectuals tasked with "thinking about the unthinkable" possibility of nuclear war (Kahn, 1962) – is a typical example of this trend (O'Brien, 2015), as is the global proliferation of explicitly "pre-emptive" responses to pressing security threats such as transnational terrorism (Stockdale, this volume). Moreover, such future-oriented approaches to societal governance have been complemented by the proliferation of broader cultural discourses advocating a more acute awareness of – and preparation for – the emergence of so-called "high-impact low-probability" events, idiomatically known as "black swans" (Taleb, 2010). For our purposes, these modes of framing contemporary sociopolitical problems are notable in that they are explicitly premised upon attempting to (re)assert a degree of control over the unfolding of the future through anticipatory mobilizations in the present. Yet the implications of these governmental trends can be problematic, as their reliance upon arbitrary imaginings of the future often privilege existing hierarchies of power and wealth while evacuating decision-making processes of their deliberative democratic character (Stockdale, 2016).

Ultimately, we believe that the uncertainties of our globalized present have made us in many ways more acutely aware of the latent dangers of an inherently unknowable future. However, we also recognize that Western societies in particular remain reluctant to modify short-termist ways of life to an extent that might change the trajectory of temporal unfolding in such a way that these dangers are forestalled. In other words, the future is too much with us, yet also in many ways absent. It has reappeared with a vengeance in an increased fear of the unknown and the resulting "risk management of everything" (Power, 2004), but it has also disappeared behind a veil of short-term profits and cynical retail politics. The emergence of this paradoxical relation to the future represents a crucial and fascinating aspect of the interaction between time and globalization, and contributions to this volume address the complex question of "global futures" from both sides.

In considering the diverse valences of time and globalization, the chapters of this volume illuminate problems as well as possibilities. Highlighting the capacity of powerful actors to harness time to extend their global dominance over human and non-human worlds, these chapters also reveal myriad instances of temporal tension, difference, and resistance. A crucial element in this book is its interdisciplinarity. Extending the complexity of its component parts, time/globalization operating together as a dynamic constellation with social, political, cultural, economic, and ecological dimensions confounds conventional epistemologies. Accordingly, the contributors to this volume deploy multiple and intersecting lenses, deriving from political science, communication studies, literature and cultural studies, anthropology and social work, to study a range of things, relations, processes and ideas. Topics covered include emerging infectious diseases, temporal sovereignty, worker exploitation and resistance, chronobiology, energy politics, the conduct of the global war on terror, activism and hope, and literary and cinematic representations of counter-temporalities. Surveying these contributions, we are struck by the richness of the account they offer of global time(s). At the same time, we cannot help but be aware of perspectives (those of the physical and health sciences, for example) that have not been included in this volume, but which would no doubt enhance the conversation. We hope that the current discussion will spark further interventions, and will draw in other participants over time, and from across the globe.

Summaries of the chapters

In his chapter entitled "Time and sovereignty in the neoliberal hegemony," Robert Hassan posits a pervasive crisis of democratic politics rooted in the changes to our relationship to space and time wrought by the universalizing logics of neoliberal globalization. Focusing on the temporal dimension of this crisis, Hassan argues that revisiting the elemental political concept of sovereignty offers a productive way to understand and critically engage with what is at stake. In this capacity, he introduces the term "temporal sovereignty" – which refers to an individual's subjective autonomy over her/his experience of time – and explores how this notion can help stimulate alternative ways of thinking to resist the processes through which time has been "expropriated and commodified" despite ultimately "*belong*[*ing*] *to us*." Combining a rich exploration of the effects that technology and acceleration exert on our apprehension of time with a meticulous consideration of the temporal foundations of sovereignty, Hassan provides a unique conceptual framework through which the political dimension(s) of globalization might be recovered, such that its homogenizing drives can be more effectively countered.

Wayne Hope's chapter, "Time, worker exploitation, and global capitalism," examines the distinctive time-related aspects of worker exploitation in a globally networked world. He identifies three epistemes of time: time reckoning (the measuring of time), temporality (the emotional, physical, and social experience of lived time), and coevalness (communication across disparate apprehensions of time). Together, they constitute a global time schema of interconnected capital-labor exploitative relationships. Such relationships are actualized at a practical level through, for example, labor and labor time arbitrage, supply chain subcontracting and precarity, and the financialization of corporate performance. Despite global capital's denial of coevalness – that is, a process of erasing the coexisting temporal perspectives associated with different peoples or cultures, the inherent conflict between capital and labor has been manifested by struggles over the capacity to use and control time in the workforce across geographies, including the potential for transnationally synchronized protests and new forms of collective temporal awareness.

Simon Orpana's chapter, "Freedom 2044? The temporality of financialization and scenario planning in *Looper*," reads the 2012 science-fiction movie *Looper* as an anxious meditation on the consequences of the financialization of the global economy. Orpana analyzes the connected themes of debt, futurity, security, and violence as they play out in a narrative that loops between 2044 and 2074, across an explicitly globalized – and Orientalized – world. Noting that *Looper* poses the solution to the problems it identifies as a nostalgic return to white American patriarchy, Orpana reads the film as exemplifying "popular culture's uncanny ability to condense nebulous, imperfectly understood social, economic and historical realities into fantasy images that are then 'magically' – which is to say, ideologically – managed and 'resolved' via narrative constructs." A critical reading of these constructs, he suggests, may help us to imagine ways to tame the turbulence and negative effects of global financial markets through new forms of global organization and solidarity.

Liam Stockdale's chapter, "Imagined futures and exceptional presents: a conceptual critique of pre-emptive security," examines the global proliferation of practices based on the imperative to govern potentially catastrophic futures through anticipatory interventions in the present. It begins with an account of how the global governance of (in)security has become thoroughly temporalized through the emergence of pre-emption as the dominant imperative of the post-9/11 era. It is then argued that pre-emption's focus on the future implies a prioritization of the imagination in the context of political decision making, which in turn enhances the discretionary subjectivity granted to governing authorities. Stockdale argues that this reveals a crucial conceptual link between pre-emptive security practices and the political exceptionalism that has been widely identified with the global war on terror, and concludes that the very notion of pre-emptive security can be seen as conceptually incoherent because this link compromises its capacity to produce a condition congruent with the understanding of "security" presupposed by its own normative framework.

Kevin K. Birth's chapter, "While the West sleeps: deglobed globalization and its consequences," examines three aspects of the "temporal logics of globalization": the dominant global time bureaucracy based on Coordinated Universal Time (UTC); the application of precision timing to high-frequency stock market trading; and the politics of the leap second debate. Birth shows how each of these protocols of time seeks at least partially to deny the importance of the actual rotation of the Earth, instead emphasizing political power and imposing a vision of global temporal uniformity. The serious consequences of this approach become visible when we realize that human circadian biology is fundamentally tied to the experience of living on a wobbly rotating planet. Highlighting the experiences of transportation workers, corporate executives, customer service agents, and Olympic athletes, Birth argues that our reliance on forms of globalized time that forget the actual globe has major destructive consequences for human health and safety.

Using the transmission in 2003 of Severe Acute Respiratory Syndrome (SARS) as a case study, Yanqiu Rachel Zhou and William D. Coleman's chapter, "Accelerated contagion: understanding the relationships among globalization, time, and disease," explores the intersections of globalization, time, and diseases – a topic that has become increasingly pertinent in the context of EIDs such as H1N1 influenza, Ebola, and Middle East Respiratory Syndrome (MERS). Questioning traditional spatiality-based approaches in the field, the authors contend that globalizing processes have changed the temporal-spatial dynamics of EIDs like SARS, in such forms as the speed of global transmission, the simultaneity of public health emergencies across geographies, and the urgent need for faster responses on an institutional level. While the technical infrastructure of global networks has provided a promising condition for accelerating surveillance and information sharing on a global level, other temporality-related challenges – such as differential capacities of the affected countries to respond simultaneously to the crisis – are yet to be tackled.

In his chapter, "The inertia of energy: pipelines and temporal politics," Brent Ryan Bellamy deploys the concept of "petroculture" – which stresses the centrality of oil to nearly all aspects of contemporary society – as a lens through which to theorize the temporalities of globalization. Beginning with a compelling discussion of existing pipeline infrastructure and its implied temporality of an "endless, oil-infused present," Bellamy proceeds to examine two diverse cases in which this petro-temporality is challenged: the proposed Mackenzie Valley Pipeline, which was aborted in the mid-1970s after vigorous contestation from Canadian indigenous groups; and the 2014 novel *The Collapse of Western Civilization*, whose narrative traces the process of global ecological collapse by looking back from an imagined future of the year 2393. For Bellamy, these two examples suggest alternative ways of envisioning our relationship to oil that might be deployed to create the sort of "political friction" required to disrupt the inertia of our present petroculture and thereby enable a different energy future.

In her chapter, "Yugoslavism: history, temporality, and the search for alternative modes of political critique," Petra Rethmann analyzes the political significance of evoking Yugoslavism and the global democratic cooperative federalism that it, and the nonaligned movement associated with it, sought to exemplify. She criticizes the views that see Yugoslavism as simply a failed political project, as a source of nostalgia, or as irretrievably lost in the past of a linear progression of time. As a contrast to the more horizontal spatial metaphors that inspire the alter-globalization movement, she suggests that Yugoslavism can act as a project of hope, which is both rooted in its post-World War II historical experience and a dream-like performative political provocation today – "a relay point between history, this present, and an untimely conviction that a better world is possible and yet to come."

Adam Barrows's chapter, "Living the rhythms of the global: time, globalization discourse, and rhythm-analysis," challenges perceptions of globalization as a process of temporal homogenization. Acknowledging currents of power that seek to enforce synchronicity, Barrows argues that time is "lived all over the globe as a negotiation of multiple and often competing rhythms." He takes up Virginia Woolf's novels as a site for exploring the emergence of a complex global time, using Henri Lefebvre's theory of rhythm-analysis and the ecological concept of panarchy to theorize its operation across scales ranging from the corporeal to the cosmological. Against uncritical celebrations of progress, or a resigned acceptance of global temporal hegemony, Barrows suggests that "globalization might more productively be understood as a perpetual negotiation between a host of competing and even mutually exclusive rhythmic patterns that constitute its global reality at any given moment and are thus always poised, if not empowered, to alter that reality."

The collection concludes with an Afterword by Sarah Sharma. This collection grew out of a series of workshops, in which a number of the contributors participated. In our final workshop, we invited Dr Sharma, whose then forthcoming book, *In the Meantime: Temporality and Cultural Politics* (Duke University Press, 2014), addressed many of the themes we were discussing, to participate as a respondent. Her Afterword, accordingly, speaks to the chapters in this volume from the perspective of her own work in Communication Studies and Cultural Studies, informed by the wider literature on time in the context of globalization.

Notes

1 Research for this chapter was funded in part by a Connection Grant from the Social Sciences and Humanities Research Council of Canada, a seed grant from the

Institute on Globalization and the Human Condition at McMaster University, and additional funding from McMaster. The authors also wish to thank Sean Pittman for research assistance, and the participants in the Workshop on Time and Globalization at McMaster University, October 19–20, 2012, for helpful insights and feedback on an earlier version of this chapter.

2 This chapter draws and expands on Huebener et al. (2015).

3 Even towering thinkers separated by vast chronological and philosophical gulfs seem to agree on this point. For instance, Heidegger once described time as inherently "unthinkable," while St Augustine famously claimed to know what time was up until the moment he was made "to explain it to one that asketh."

4 Bertrand Lançon notes, for instance, that timekeeping in the Roman Empire involved complex negotiations between different localities and religious customs. "The term 'Roman calendar'," he writes, "must not mislead us. It was not a calendar that would have been followed throughout the empire; on the contrary, there was a multiplicity of calendars, often peculiar to individual cities, such as that observed at Alexandria" (Lançon, 2000, p. 131). Methods for counting the hours of the day also differed between Christian and non-Christian Romans (Lançon, 2000, p. 132).

References

Adam, B. (1994) Perceptions of Time. In Tim Ingold (ed.) *Companion Encyclopedia of Anthropology: Humanity, Culture and Social Life* (pp. 503–526). London: Routledge.

Adam, B. (1998) *Timescapes of Modernity: The Environment and Invisible Hazards.* London: Routledge.

Adam, B. (2004) *Time.* Cambridge: Polity Press.

Allen, T. (2008) *A Republic in Time: Temporality and Social Imagination in Nineteenth-Century America.* Chapel Hill: University of North Carolina Press.

Anderson, B. (1991) *Imagined Communities: Reflections on the Origins and Spread of Nationalism*, 2nd edn. New York: Verso.

Anderson, B. (2010) Pre-emption, Precaution, Preparedness: Anticipatory Action and Future Geographies. *Progress in Human Geography*, 34(6), pp. 777–798.

Barrows, A. (2011) *The Cosmic Time of Empire: Modern Britain and World Literature.* Berkeley: University of California Press.

Bauman, Z. (2000a) Time and Space Reunited. *Time and Society*, 9(2–3), pp. 171–185.

Bauman, Z. (2000b) *Liquid Modernity.* Cambridge: Polity Press.

Beck, U. (1999) *World Risk Society.* Cambridge: Polity Press.

Beck, U. (2002) The Terrorist Threat: World Risk Society Revisited. *Theory, Culture & Society*, 19(4), pp. 39–55.

Blaise, C. (2001) *Time Lord.* Toronto: Vintage Canada.

Bloch, E. (1991) *Heritage of Our Time.* Cambridge: Polity Press.

Castells, M. (1997) *The Power of Identity: The Information Age – Economy, Society, and Culture, Volume II*, 2nd edn. Oxford: Wiley-Blackwell.

Cazdyn, E. (2012) *The Already Dead.* Duke University Press.

Cazdyn, E. and Szeman, I. (2012) *After Globalization.* Chichester: Wiley-Blackwell.

Chakrabarty, D. (2009) The Climate of History: Four Theses. *Critical Inquiry*, 35, pp. 197–222.

Chomsky, N. (2000) *Globalization: The New Face of Capitalism. Democracy Now*, February 3. www.democracynow.org/2000/2/3/noam_chomsky_globalization_the_new_face#transcript.

Coelho, K. (2003) Timed Out: Temporal Struggles between the State and the Poor in the Context of U.S. Welfare Reform. *Arizona Anthropologist*, 15, pp. 72–98.

Conway, J. and Singh, J. (2009) Is the World Social Forum a Transnational Public Sphere? Nancy Fraser, Critical Theory, and the Containment of Radical Possibility. *Theory, Culture & Society*, 26(5), pp. 61–84.

Crang, M. (2007) Speed=Distance/Time: Chronotopographies of Action. In Hassan, R. and Purser, R.E. (eds) *24/7: Time and Temporality in the Network Society* (pp. 62–89). Stanford, CA: Stanford Business Books.

Crang, M., Crosbie, T. and Graham, S. (2007) Technology, Time-space, and the Remediation of Neighbourhood Life. *Environment and Planning A*, 39, pp. 2405–2422.

Crary, J. (2013) *24/7: Late Capitalism and the Ends of Sleep*. London and New York: Verso.

der Derian, J. (1990) The (S)pace of International Relations: Simulation, Surveillance, and Speed. *International Studies Quarterly*, 34(3), pp. 295–310.

Dewdney, C. (2008) *Soul of the World: Unlocking the Secrets of Time*. Toronto: HarperCollins.

Fabian, J. (1983) *Time and the Other: How Anthropology Makes its Object*. New York: Columbia University Press.

Felski, R. (2000) Introduction: Timely Meditations. In *Doing Time: Feminist Theory and Postmodern Culture* (pp. 1–32). New York University Press.

Galison, P. (2003) *Einstein's Clocks, Poincaré's Maps: Empires of Time*. New York: W. W. Norton.

Giddens, A. (1999) Risk and Responsibility. *Modern Law Review*, 62(1), pp. 1–10.

Gupta, A. (1992) The Reincarnation of Souls and the Rebirth of Commodities: Representations of Time in "East" and "West." *Cultural Critique*, 22, pp. 187–211.

Hansen, M.B.N. (2009) Living (with) Technical Time: From Media Surrogacy to Distributed Cognition. *Theory, Culture and Society*, 26(2–3), pp. 294–315.

Harvey, D. (1992) *The Condition of Postmodernity: An Enquiry into the Origins of Cultural Change*. Oxford: Blackwell.

Hassan, R. (2005) Timescapes of the Network Society. *Fast Capitalism*, 1(1). www.uta.edu/huma/agger/fastcapitalism/1_1/hassan.html.

Hassan, R. and Purser, R. (eds) (2007) *24/7: Time and Temporality in the Network Society*. Stanford, CA: Stanford University Press.

Hochschild, A.R. (2005) On the Edge of the Time Bind: Time and Market Culture. *Social Research*, 72(2), pp. 339–354, 504.

Hope, W. (2009) Conflicting Temporalities: State, Nation, Economy and Democracy under Global Capitalism. *Time & Society*, 18(1), pp. 62–85.

Huebener, P., O'Brien, S., Porter, T., Stockdale, L. and Zhou, R. (2015) Introduction: Exploring the Intersection of Time and Globalization. *Globalizations*, 13(3). DOI: 10.1080/14747731.2015.1057046.

Kahn, H. (1962) *Thinking about the Unthinkable*. London: Weidenfeld & Nicolson.

Lançon, B. (2000) *Rome in Late Antiquity: Everyday Life and Urban Change, AD 312–609*. (Antonia Nevill, trans.). New York: Routledge.

Latour, B. (1990) Technology is Society made Durable. *The Sociological Review*, 38(S1), pp. 103–131.

Law, J. (1986) On the Methods of Long-Distance Control: Vessels, Navigation and the Portuguese Route to India. In John Law (ed.), *Power, Action and Belief: A New Sociology of Knowledge?* (pp. 234–263). London: Routledge and Kegan Paul.

LeBel, S. (2016) Fast Machines, Slow Violence: ICTs, Planned Obsolescence, and e-Waste. *Globalizations*, 13(3), pp. 300–309.

Leccardi, C. (2007) New Temporal Perspectives in the "High-Speed Society." In Hassan, R. and Purser, R.E. (eds) *24/7: Time and Temporality in the Network Society* (pp. 25–37). Stanford, CA: Stanford Business Books.

Lefebvre, H. (2004) *Rhythmanalysis*. London: Continuum.

Levine, R. (1997) *A Geography of Time: The Temporal Misadventures of a Social Psychologist, or How Every Culture Keeps Time Just a Little Bit Differently*. New York: Basic Books.

Lewis, M. (2014) *Flash Boys: A Wall Street Revolt*. New York: W.W. Norton.

Li, V. (2015) The Untimely in Globalization's Time: Don DeLillo's Cosmopolis. *Globalizations*. DOI: 10.1080/14747731.2015.1056493.

Lim, B.C. (2009) *Translating Time: Cinema, the Fantastic, and Temporal Critique*. Durham, NC: Duke University Press.

Lübbe, H. (2009) The Contraction of the Present. In Harmut Rosa, H. and Scheuerman, W.E. (eds), *High-Speed Society: Social Acceleration, Power, and Modernity* (pp. 159–178). University Park, PA: Penn State University Press.

May, J. and Thrift, N. (2001) Introduction. In May, J. and Thrift, N. (eds), *Timespace: Geographies of Temporality* (pp. 1–46). London: Routledge.

McCallum, E.L. and Tuhkanen, M. (2011) *Queer Times, Queer Becomings*. Albany: SUNY Press.

McQuire, S. (1999) Blinded by the (Speed of) Light. *Theory, Culture and Society*, 16(5–6), pp. 143–159.

Mittal, E. and Raghunandan, Y. (2012) *presence* (video). Behind the Tin Sheets. Screened at the Workshop on Time and Globalization, October 19, Workers Arts and Heritage Centre, Hamilton, Canada. See also www.tinsheets.in/context/presence/.

Morris, M. (1998) Introduction: History in Cultural Studies. In *Too Soon, Too Late: History in Popular Culture* (pp. 1–30). Bloomington: Indiana University Press.

Nanni, G. (2013) *The Colonisation of Time: Ritual, Routine and Resistance in the British Empire*. Manchester: Manchester University Press.

O'Brien, S. (2015) "We Thought the World was Makeable": Scenario Planning and Postcolonial Fiction. *Globalizations*, 13(3).

Ong, A. (1987) *Spirits of Resistance and Capitalist Discipline: Factory Women in Malaysia*. Albany: SUNY Press.

Paris, M. (2012) Budget Shortens Environmental Review Process. *CBC News*, March 29. www.cbc.ca/news/politics/budget-shortens-environmental-review-process-1.1270727.

Parkins, W. and Craig, G. (2006) *Slow Living*. London: Berg.

Petrick, K. (2015) Strategic Planning in the "Empire of Speed." *Globalizations*. DOI: 10.1080/14747731.2015.1056496.

Porter, T. and Stockdale, L. (2015) The Strategic Manipulation of Transnational Temporalities. *Globalizations*. DOI: 10.1080/14747731.2015.1056497.

Power, M. (2004) *The Risk Management of Everything: Rethinking the Politics of Uncertainty*. London: Demos.

Ricoeur, P. (1984) *Time and Narrative*, Vol. 1. (Kathleen McLaughlin and David Pellauer, trans.). Chicago, IL: University of Chicago Press.

Rifkin, J. (1989) *Time Wars: The Primary Conflict in Human History*. New York: Simon & Schuster.

Roller, D.W. (2010) *Eratosthenes' Geography*. Princeton, NJ: Princeton University Press.

Rosa, H. (2009) Social Acceleration: Ethical and Political Consequences of a Desynchronized High-Speed Society. In Rosa, H. and Scheuerman, W. (eds) *High-Speed Society: Social Acceleration, Power, and Modernity* (pp. 77–112). University Park, PA: Pennsylvania State University Press.

Rosa, H. (2013) *Social Acceleration: A New Theory of Modernity.* In Trejo-Mathys, J. (trans.). New York: Columbia University Press.

Rosa, H. and Scheuerman, W.E. (eds) (2009) *High-Speed Society: Social Acceleration, Power, and Modernity.* University Park, PA: Penn State University Press.

Rutherford, J. and Davison, S. (eds) (2012) *The Neoliberal Crisis.* Lawrence Wishart.

Sassen, S. (2000) The Global City: The De-nationalizing of Time and Space. In Ockman, J., *The Pragmatist Imagination: Thinking about "Things in the Making"* (pp. 254–265). Princeton, NJ: Princeton Architectural Press.

Scheuerman, W.E. (2009) Citizenship and Speed. In Rosa, H. and. Scheuerman, W.E. (eds) *High-Speed Society: Social Acceleration, Power, and Modernity* (pp. 287–306). University Park, PA: Penn State University Press.

Sharma, S. (2014) *In the Meantime: Temporality and Cultural Politics.* Durham, NC and London: Duke University Press.

Sioui, G. (2008) *Histoires de Kanatha Vues et Contées – Histories of Kanatha Seen and Told.* Ottawa: University of Ottawa Press.

Sobel, D. (1995) *Longitude.* New York: Walker and Co.

Stiegler, B. (1998) *Technics and Time, 1. The Fault of Epimetheus.* (Richard Beardsworth and George Collins, trans.). Palo Alto, CA: Stanford University Press.

Stockdale, L. (2013) Imagined Futures and Exceptional Presents: A Conceptual Critique of "Pre-emptive Security." *Global Change, Peace & Security*, 25(1), pp. 141–157.

Stockdale, L. (2016) *Taming an Uncertain Future: Temporality, Sovereignty, and the Politics of Anticipatory Governance.* Lanham, MD: Rowman & Littlefield International.

Taleb, N.N. (2010) *The Black Swan: The Impact of the Highly Improbable.* New York: Random House.

Thrift, N. (2004) Remembering the Technological Unconscious by Foregrounding Knowledges of Position. *Environment and Planning D*, 22, pp. 175–190.

Urry, J. (2009) Speeding Up and Slowing Down. In Rosa, H. and Scheuerman, W.E. (eds) *High-Speed Society: Social Acceleration, Power, and Modernity* (pp. 179–198). University Park, PA: Pennsylvania State University Press.

Virilio, P. (interviewee) and Armitage, J. (interviewer). (2000) CTheory Interview with Paul Virilio (transcript interview) (Patrice Riemens, trans.), October 18. www.ctheory.net/articles.aspx?id=132.

Zhou, Y.R. (2015) Time, Space and Care: Rethinking Transnational Care from a Temporal Perspective. *Time & Society*, 24(2), pp. 163–182.

2 Time and sovereignty in the neoliberal hegemony

Robert Hassan

Introduction: banal processes and their interesting consequences

In a 2008 essay titled "The End of Neoliberalism?" economist Joseph Stiglitz strained to put a positive spin on the future prospects for a global economy that was then sliding toward what looked to be a horrible precipice. He wrote:

> Neoliberal market fundamentalism was always a political doctrine serving certain interests. It was never supported by economic theory. Nor, it should now be clear, is it supported by historical experience. Learning this lesson may be the silver lining in the cloud now hanging over the global economy.
>
> (Stiglitz, 2008)

The hoped-for "lesson" was to be that the severity of the crisis would shake up the political process and its institutions, causing them to realize, through rational debate and a shared interest, that common sense would, or at least could, prevail over the overweening "political doctrine" of neoliberalism.

Stiglitz implies that the phenomenon and consequences of neoliberal globalization stem from politics and political consciousness. I would agree. However, he further assumes that intact and properly functioning democratic structures and processes are available to help generate the common-sense and common-interest attitudes that would be able to draw the correct lessons from the crisis – and then put things to rights. The argument is rather like that of Francis Fukuyama, who in his 1989 "The End of History?" and in subsequent writings, sees liberal democracy of a distinctly conservative kind as a natural default position for the modern era; an "essence" of Enlightenment political institutions and cultures that we can reach to, and which will remind us of the reasoned approach to take to (any) crisis. In short, Stiglitz laments the hegemony of neoliberalism, but imagines that some kind of restitution can be made through recourse to reasoned and rational liberal democratic structures.

The hegemony remains, however, and the lack of any tangible lessons learned in the years since 2008 points to the fact that neoliberalism is not just hegemonic, but that countervailing pressures are not strong enough and organized enough to make any difference to fundamentalist free-market "political doctrine." Neoliberalism's success in creating its own surface reality whilst obscuring a deeper reality of crisis means, amongst other things, that respected and influential thinkers such as Stiglitz who want a fairer and more democratic world, continue to imagine that we still have relatively undamaged political institutions and political tools to guide us; a "silver lining" amongst the present wreckage – if only we would reach out to rediscover and use them.

To counter such insouciance I begin by articulating what I see as a double banality that lies at the heart of the present neoliberal global hegemony. It is a twofold, or two-dimensional commonplace that, through its very ordinariness, blinds us to its more consequential importance. This is (or these are) *globalization itself* and the *crisis of politics* that enables it. That today often they invoke a yawn of familiarity, or a slightly more engaged feeling that nothing new can possibly be said about them, is testament to their power as tranquilizing media tropes which we too often allow to function and run their course with little sustained reflection. We do so in part because the world that they create obscures a deeper reality of their logic and the consequences of their unfolding. Nevertheless this double banality conceals a double crisis – or what could be, or in fact more correctly should be, a singular crisis of space and time. However, the "unhelpful dualism" (Thrift and May, 2002, p. 1) of thinking of space and time as separate categories, and which pervades much theorizing of these still, has a certain explanatory value that I will retain briefly in order to examine the dimensions of crises in our post-modern context. I do so to more clearly link them back in such a way that lays the groundwork for the main argument that I want to make here.

Before defining that argument, I want to look first at the spatial and temporal dimensions in their individual distinctiveness. The crisis of space today is one deriving from the particular logic of neoliberal globalization. However, the roots of this crisis cannot be adequately perceived or understood from the perspective of what is termed, broadly, as geopolitics, politics that carry conceptions of territorial space or the nation-state as the constitutive element for the study and practice of global politics (Ó Tuathail et al., 2006). Traditional methodologies no longer work so durably because neoliberalism's border-free logic has visited its creative destruction upon the psychic and physical borders of individuals and communities and polities in every part of the planet since the 1970s. Moreover, the contradiction between the free movement of information and capital and the much more restricted traffic of humanity continues to engender strife and disharmony within and between every continent.

Thinking on the temporal crisis of the political process is rather less conventional in its articulation. The temporal dimension of the political process has only recently come to some salience in academia, and the rubric "chronopolitics," "a term descriptive of the relation of time-perspectives to political decision-making" (Wallis, 1970, p. 102) which first emerged several decades ago, captures some of the thinking and the debates. Broadly speaking, chronopolitics seeks to make time a more prominent factor in respect of how our politics works and how it evolves in the liberal democratic sphere.

Two main approaches are recognized here. One is of time as a broad determiner, where the Newtonian derived chronology of clock and calendar and of an unfolding historical process gave to modern politics its rhythms, its linearities, its sequences and its familiarities (Hutchings, 2008, pp. 34–35; Adam, 2004, p. 139). The other is to see time act as either enabler or constrainer in the political process – and in both of these ways to the detriment of democracy. For example, time structures that accelerate, serve to enable executive power within liberal democracy (Scheuerman, 2004, pp. 27–70). This is because in the increasingly global context for politics, the need for rapid decision making, especially in economic questions, enables "executive discretionary powers" (Scheuerman, 2004, p. 37) to bypass a slow-moving legislature and to concentrate power at the pinnacle of government. Time also constrains the political process, again through the context of its acceleration. Paul Virilio, for example, argues that machine-driven acceleration, especially in our computer-information age, restricts people at the level of the experience of everyday life. Here, the speed-up can disorientate, preventing the individual from being fully aware of and able to act upon her civic rights and duties, rendering her vulnerable to what Virilio describes as a "dictatorship of speed" (Virilio, 1995).

There is more recent work that seeks to merge geopolitics with chronopolitics. We see this in the work of Ian Klinke, who coins the rather ungainly term "geochronopolitics" (Klinke, 2013, p. 678). He argues that to understand politics fully, especially in our age of globalization, concepts of space and time must transcend Thrift and May's problematic of the "unhelpful dualism," and be merged. However, the fusion, coming from the geographer Klinke, is an odd one. It is through the development of literary theorist Mikhail Bakhtin's idea of the "chronotope" (meaning time space), that these are understood best at the discursive level of language and through the narratives it produces. I say "odd" not in any disrespectful sense, but because if we wish to think about politics, about globalization, about space and time, then we need to think as *deeply politically* as we can, because these processes have all been conditioned by political choices. Although the approaches I have too briefly covered here have value, what they lack is applicability to politics as it evolves in our post-modernity. That is to say, our analysis needs to be able to link political conceptions of space and time in a manner that reflects the way that neoliberal globalization is universalizing in its logic; to be cognizant, in other words, of the idea that what globalization does to our relationships with space and time it does also to the political process.

What I will argue in this chapter is that it is possible to get to the roots of the crisis of space and time within the grip of neoliberal globalization through the political and universal concept of *sovereignty*. This is sovereignty in respect of autonomy at the level of the individual, but it also connects to the praxis of sovereignty as an organizing mechanism for politics at the register of economy and society. I see these conceptions as being based upon traditional theorizations from political theory, but here I wish to place them upon a continuum, between two poles of sovereignty that links the private to the public, and the individual to the social. This more flexible and fluid framework is necessary, I believe, because as we will see, it helps capture the transformed nature of politics, especially in respect of time and temporality in the globalizing context.

The particular idea of sovereignty that I will introduce is that of "temporal sovereignty." It is a form of political sovereignty that is equal to and coextensive with the more familiar idea of spatial sovereignty. However, temporal sovereignty is missing in our current political lexicon – for reasons, as I will show, that have been largely forgotten in the literature on political sovereignty that began with Thomas Hobbes. Moreover, what we are dealing with is much more than the processes of "political decision making" in time, but encompassing the deeper philosophical questions of the subjective modern political condition that began with Hobbes and was articulated more fully during the Enlightenment.

The essential lop-sidedness in our thinking on sovereignty, which focuses on space (and the body) as opposed to a combined theorization of the spatial and the temporal, is only now having effects in political conduct. What David Harvey (1989) called "time-space compression" has fundamentally changed our relationship with time and space. Neoliberal globalization changes the relationship from that within a machine and clock and analogue-shaped world, to a computer and networked and digital reality, where the anchors of our modernity are being shaken loose. In this highly fluxual landscape, a crisis for politics has emerged (Bauman, 2000, p. 108). It is a crisis of temporality, but not in the sense with which we are familiar in current time studies where the Scheuerman and Virilio perspectives act as (to reuse an earlier phrase) representative poles along a fairly wide, but nonetheless shallow, continuum. The idea of a crisis of temporal sovereignty argues that *time belongs to us* as individuals and as social beings, and constitutes the very basis upon which the theory and practice of democracy rests. However it is this lived time and social time that is being expropriated and commodified through our digital existence in ways we could not have conceived of even a generation ago. Temporal sovereignty, moreover, is key to understanding the limits to and crisis of our more familiar concept of spatial sovereignty. Ultimately, then, recognition of the fuller spatio-temporal dimensions of political sovereignty will give a fresh perspective on neoliberal globalization and suggest some ways forward from the political impasse that post-modern capitalism creates for us today. However, let us first consider the spatial crisis – the better to understand the need for the recognition of our temporal rights as political beings.

Expanding globalization, contracting sovereignty

Globalization as an economic and social condition was not recognized as such in previous iterations of capitalist expansion. The concept of physical space was not salient or problematic during the phase of 19th-century colonialism, for example, because there seemed a surfeit of it. Unproblematic until the world was carved up by the Great Powers, thereby creating the tensions that led to the world wars of the 20th century. Even after 1945, the war's devastation – which had been general in Europe and Japan – created much "new" space into which capital (primarily from the United States) could flow in the rebuilding of the post-war world. This 30-odd-year phase of global economic boom, the so-called "Golden Age," pushed issues of space and of political sovereignty to the background in the context of what was essentially nation-state-led recovery, expansion and consolidation under a broadly social democratic political ideology (Judt, 2005, pp. 360–386).

The immanent process of capitalist expansion became connected to the post-modern concept of globalization only when the *finite* space of the planet became a restraining factor upon capital's need to grow infinitely. The period of inter-national expansion of capitalism from 1945 until the mid-1970s brought to the fore the idea of the finiteness of territorial space, though only at the end of the long boom. By the 1970s capital could find fewer outlets for its "realization" as profit (Harvey, 1983, p. 192) principally because in the non-communist world, fresh territories for the geographical displacement of capital were shrinking. David Harvey outlined this process with some prescience in his 1983 book *Limits to Capital*, and in his 1989 *Condition of Postmodernity* he introduced the concept of "time-space compression" (Harvey, 1989, pp. 240–242) to describe the nature of the growing crisis for globalizing capitalism. With the contraction of new space for the realization of profit, capitalism entered a period of crises in the 1970s, the "resolution" of which would become Harvey's "postmodernity" and where "flexible accumulation" (Harvey, 1989, pp. 147–197) would become capital's new modality. This approach was a political choice – an ideology we now call neoliberalism – and this neoliberalism and the "time-space compression" that it would enable would not have been possible without the revolution in information technologies. We will take up this particular trajectory and its temporal implications in the next section, but to continue, globalization, in Harvey's reading, is another term for what he calls a "geographical and geopolitical crisis" (Harvey, 1989, p. 186) from which emerges another element of crisis – which is that of sovereignty.

The idea of sovereignty as a central component of modern political theory comes to us from Thomas Hobbes. His 1651 *Leviathan* systematized sovereignty as a juridical-political idea. Hobbes located sovereignty as an instrument of power invested in the person of the monarch. *Leviathan* also introduced the concept of the social contract, where the individual would abdicate or "lay down his rights to all things" and transfer them to the invested ruler who would be charged to use this power to protect people from an otherwise chaotic society (Hobbes, 2011, p. 16). Sovereignty was embodied and individualized, but as the 1648 Treaty of Westphalia instituted, it was also territorial and was "contained" within the space to where political power's writ extended.

Within this objective territorial concept was the more democratic idea of individual and subjective sovereignty, one based upon property and ownership that gave another dimension to Hobbes's concept of the social contract between ruler and ruled. This was a powerful dynamic containing the social and economic basis for capitalist modernity. As Andro Linklater wrote of the early 1500s in his book *Owning the Earth*:

> The idea of individual, exclusive ownership, not just of what can be carried or occupied, but of the immovable, near-eternal earth, has proved to be the most destructive and creative cultural force in written history. It has eliminated ancient civilisations wherever it has encountered them, and displaced entire peoples from their homelands, but it has also spread an undreamed-of degree of personal freedom and protected it with democratic institutions wherever it has taken hold.
>
> (Linklater, 2014, p. 5)

Linklater indicates here that the ownership of space and the declaring of it sovereign is fundamental in the creation of bourgeois democracy. Flowing from this – and in a way that again reaches back to Hobbes to universalize his idea of the sovereign body – is that people *as individuals* began to enjoy sovereign political rights articulated and institutionalized through Enlightenment thought. The rights that were "laid down" in Hobbes were picked up again by thinkers such as John Locke and deemed to be essential and inalienable in all people.[1]

Tied as they were to the interrelated dynamics of politics and expansionary capitalist economics, territorial and individual sovereignty within the project of modernity was always going to be problematic at some point; we just acted as if it were not. The fateful decade of the 1970s changed that. The "crisis of space" that led to "globalization" created a crisis of sovereignty of a kind that had not been seen – in its generalizing of certain of its characteristics – within capitalist democracy. A pre-eminent characteristic was that for the first time inter-dependent circuits of global capital were de-linking from national domiciles in response to the spatial and temporal pressures and imperatives captured in the often-repeated call by politicians, economists and chief executives, that capital now needed to be *flexible* – a term connoting both movement (in space) and speed (in time), a logic and ideology that Harvey encapsulates in his "flexible accumulation."

Notwithstanding the relative burgeoning of new states (55 since 1974), and the various assertions of sovereign power in, for example, border politics across the world (see Vaughan-Williams, 2009, for a critique of such reassertions), territorial sovereignty is now deeply compromised by the far more powerful and complex and integrated networked flows of global capital. Nation-states, from the United States down are increasingly restricted in their autonomy and sovereignty, especially in the economic realm. Contrast this with a Goldman Sachs or Deutsche Bank, which have amassed levels of unprecedented economic and political power (Cohan, 2011, pp. 302–314). Global corporations have intensified their leverage over the affairs of national governments in inverse proportion to the decrease in civic freedoms enjoyed by elected governments and those they represent. Moreover, the diminution of nation-state sovereign power by borderless economic flows is power that has been *abrogated*, not forcibly taken away. Wendy Brown has considered "political sovereignty's [historical] bid to subordinate the economic to the political," and writes:

> This is not to say that the modern sovereign nation-state always heavy-handedly governs or regulates the economy but that it is premised on the notion that it can, and that it *decides* whether and when to do so. Laissez-faire capitalism is as much the expression of such a decision as state socialism is.
>
> (Brown, 2010, p. 57, emphasis in original)

The decision to adopt *laissez-faire* capitalism may reduce sovereignty at the level of territorial writ, but it is also an impoverishment of sovereignty at the level of the individual where macro-economic forces conspire to discard the rights and protections inherent in the social contract. Perhaps the most dispiriting illustration of this at the level of everyday life is the record number of refugees (51 million in 2014) unable to avail themselves of the sovereign protection of a state. These millions are invisible to most of us in the rich economies, except as a spectral or existential threat. Also hidden for the most part are the global homeless, a fluxual number across the planet, but said to number up to 100 million even prior to the global financial crisis of 2007 (Kothari, 2005). Having no access to a secure place of shelter violates the personal freedoms and individual sovereignty codified in the International Covenant on Civil and Political Rights, where the rights of these many millions to vote, to privacy, to freedom of expression and so on become drastically curtailed.

The nation-state's abrogation of its political responsibilities has had global consequences. Perry Anderson has noted the lack of plausible political alternatives to neoliberal rule, arguing that our global monopolitical culture has no precedent (Anderson, 2000, p. 17). Neoliberalism has no rival and therefore no counter-tendency to justify its conduct to. This global problem of course has its local articulations manifest in a detached political class whose behavior sows more apathy than anger in people. This is captured in its cynical essence in French President Jacques Chirac's comment on the French voters' decision in 2005 to reject the European Constitution: "You have rejected the European constitution by a majority. It is your sovereign decision and I take note of it" (Brown, 2010, p. 250). This meant ignoring the "sovereign decision" taken by popular vote and colluding with an elite group of European politicians to get the kind of treaty that would be acceptable to his class – which was then signed by them into law as the Lisbon Treaty in 2007.

Neoliberalism has been a stunning victory for capitalism. The dynamics of space, of sovereignty and the political process are being emptied of their former content and replaced by the logic of accumulation. As Brown argues, nation-states still have political choice vis-à-vis the processes of capitalist globalization, but lack the *political will* to decide to act upon their historical responsibilities to their people. Political will, indeed, has been turned negatively toward the great mass of people in the liberal democracies as a direct consequence of the individualism that neoliberal ideology has promoted. If the sovereign power of the state is delimited through neoliberal globalization, it is still sufficient to be exerted over those very elements of society that have been most weakened by globalization. So people, as "individuals" living in an individualistic world, find themselves increasingly isolated and exposed to a state power that can harass and vilify and scapegoat victims *for being victims.* Such groups, which often comprise a growing underclass, can be pursued as elements of what Margaret Thatcher once described (speaking of the British mineworkers) as the "enemy within" and therefore the real cause of all kinds of social malaise (Milne, 2004). Gertrude Himmelfarb has noted the flip side of the neoliberal fetishizing of the individual, where "as the culture becomes more aggressively individualistic, so the state becomes more aggressively expansive" (Himmelfarb, 2012). This expansion intrudes into the spaces of the lives of people, especially the most vulnerable, those on welfare, those who hold political opinions not congruent with free-market orthodoxy, those seeking asylum, and so on. Somewhat more dramatically, the deepening and widening extent of the secret state's mistrust of its own people was exposed by whistleblower Edward Snowden in 2013 (Harding, 2014).

In many ways, neoliberalism and the globalization it has promoted is unique: it is a political philosophy that is concerned with the economy above all else. The drive for the free flowing of capital has created a corresponding contraction of the sovereignty of the nation-state and the individual sovereignty of the overwhelming majority of humanity, the 99 percent that the Occupy Movement identifies as capitalism's victims. However, sovereignty has a temporal aspect as well, and recognition of this is vital if we are to have any prospect of reviving and refashioning politics for the networked age. Before moving to that discussion, I want first to preface it with a look at the nature of time in its technological (computerized) context.

Temporal objects and diminished subjects

As a professional geographer it was perhaps understandable that in his influential theorizing of "time-space compression" in *The Condition of Postmodernity*, David Harvey concentrated almost wholly on space. Nonetheless, the "crisis of space" that he hypothesized so wonderfully was also a *crisis of time* – and perhaps more so. The idea that neoliberal globalization has brought with it a new relationship to time is now fairly widespread. Broadly speaking, such thinking considers the relationship as one of "social acceleration" (Scheuerman, 2004; Tomlinson, 2007; Rosa, 2013), where life quickens, where the logic of speed is pervasive, and where "unoccupied time is vanishing" (Gleick, 2000, p. 10).

These and others make valuable contributions to our understanding of speed and post-modern life, but apart from Hartmut Rosa in his 2013 book *Social Acceleration*, none *problematizes time* – that is to say, to look at the nature of time *itself* as a social/individual phenomenon, as opposed to mainly dwelling upon the multitudinous *expressions* of its acceleration. Rosa himself, as a sociologist drawing expertly upon phenomenology to define temporality, is nonetheless content to devote his book to observing and defining with his sociologist's eye, the consequences of acceleration on phenomenological time. His "political" discussion is equally descriptive, with the book ending feebly with an attempt at a "redefinition of modernity" with which to make sense of the present (Rosa, 2013, pp. 277–299). Rosa's work has benefit in that it problematizes time and sees it as something originating in the experience of the individual, and as a process that may be shared socially between individuals as the social time that he sees as accelerating. However, such passive spectatorship does not get us far if we seek solutions to the dual crises of space and time that come to us from the colonizing logic of neoliberal globalization. The problem is political; and the solution, as Brown reminds us, is political also. Central to thinking about the problem, and to be able to do something about it, is not a "redefinition of modernity" – an epoch that we have already transcended – but a new understanding of sovereignty, one that views it as both spatial *and* temporal. Thus we need to think through our relationship with time in the context of neoliberal globalization. More, we need to think in the context of the technological processes that made this globalization possible – information technology.

Since the 1980s the determining context of our post-modern and neoliberal lives has been technological – or more precisely the computerization of economy, culture and society. The contemporary glories of Wall Street, of Silicon Valley and its plutocrats, of eBay, Amazon, Apple, Facebook, the app economy and digital capitalism *tout court*, are what bring neoliberalism to life – and ever more deeply into our lives. To become mobile and networked nodes by means of ever cheaper and more powerful computer devices is the very basis of globalization, of space time compression, and of social acceleration. In short: to be networked and increasingly to inhabit digital space as opposed to physical place, changes our relationship with time and constitutes a global timescape.

What I term "network time" (Hassan, 2003, 2009) is the basis for and expression of this new post-modern time. Network time is the context of the time that we produce and respond to in our networked lives. It is the time of connectivity, where the once-supervening time of the clock is no longer so powerful and determining of the rhythms of the day. The context of this time is as diverse and variable as the possible number of contexts that people can create when they connect one-to-one or one-to-many to work, collaborate, share, produce, entertain, consume and so on. Within the digital sphere it matters less in which clock time zone we are physically placed and more the nature (quality, strength, speed, duration, etc.) of the connection itself – which is a technological issue. Data flows may be slow or fast, or somewhere in between; Wi-Fi signals may be patchy or constant; video connections may drop out or freeze, or be just like watching television. These and other variables influence network time and our experience of it, but one thing is certain in this technological context: the speed of the network is always oriented towards open-ended acceleration, a speed limited only by levels of technological sophistication, and competition ensures that the network gets faster and more sophisticated every day. What is revolutionary (and post-modern) about this production and experience of time is that the clock-time rhythm of modernity and the clock's *basso continuo* role in modern life have been undermined, displaced and increasingly replaced by network-generated temporal flux, temporal contingency and temporal acceleration.

Potentially, network time could free us from the predictable and commodified rhythms of modern, industrial clock time. It could constitute a positive timescape wherein people acting as more genuinely temporal beings could have more temporal autonomy in their lives. However, the network is a profoundly commodified space, one that would not exist without the imperatives of profit and competition. Competition compels the network to colonize increasing registers of economy, culture and society, encompassing, as it does so, increasing elements of one's own life within the digital space to become "always on" and always available. Notwithstanding the potentially limitless diversity of network time contexts, the flattening logic of the marketplace ensures, as Jonathan Crary notes, that "the vast majority of these micro-worlds, despite their patently different content, have a monotonous sameness in their temporal patterns and segmentations" (Crary, 2014, p. 54).

Computing not only produces the context for time in our post-modernity, but also mass-produces the technologies of time that help shape our experience of the network. Bernard Stiegler has written insightfully about the nature of time and its technological production. He introduces the term "temporal objects" to describe how time becomes compressed and commodified and objectified in our networked lives. Drawing upon Husserl's phenomenology, Stiegler argues that in the Internet-globalization age, the saturation of media commodities such as film, music, television, videos, social media, etc., constitute a new relationship with time and transform how we distinguish and perceive past, present and future. For Stiegler the pervasion of these real-time media products within the flows of the network disengages us from the perception of time, of *différance* in lived time, between what Husserl terms "primary, secondary, and tertiary retention" (Stiegler, 1998, p. 17). The "temporal objects" that constitute the form and shape of our networked lives are responsible for what Stiegler argues to be the synchronization of time consciousness with the temporality of the objects themselves (Hansen, 2004). Time is thus constituted in a digital way that has radically different effects from that of the clock. Through connectivity, synchrony with the network "objectifies" time from human experience in an equally radical way. For Stiegler, this is "the expropriation of human time submitted to commodity-time" (Stiegler, 2011, p. 27)

Stiegler's insight is useful. However, it was conceived when the Internet and the network were in their infancy – and the "expropriation" he speaks of later extends still further than he credits. The digital network and the temporal objects with which it changes human time consciousness are now extensively and intensively developed. Today, there is one smartphone for every five people, and the ratio is rapidly narrowing (Heggestuen, 2013). This is to say nothing of the massive growth of tablet, laptop, desktop and wireless connectivity. The experience of lived time as we have known it since the Industrial Revolution, with its unfolding of past, present and future, is being supplanted by a present-centered form of time reckoning, a network-generated time of flux and immediacy leading to the inevitable perception and experience of acceleration. The objectification of time through technical objects transforms time consciousness (our subjective time consciousness) to an extent that the clock could never have attained. The colonization of our subjectivity by temporal objects is refashioning us as post-modern *temporal subjects* – individuals who are vulnerable to a network time dependency that leaves little time, subjectively speaking, for reflection or critique or sustained concentration. This network time is becoming so deeply part of our temporal experience that we become *diminished* temporal subjects through the ascendancy of digital temporal objects.

All this may read pessimistically, somewhat like Herbert Marcuse's baleful analysis of late-modern capitalism as "total administration" where technological society transforms its subjects into "stupefied" and trapped "objects of total administration" (Marcuse, 1990, p. 7). This is indeed the direction of rationalized technological development. However, human subjects are irreducibly *political*, and as Mark Hansen says of Stiegler's analysis, "it obtains a crucial ethical dimension [that] will allow us to retain a human perspective, *even when the time of the human seems to be eclipsed by realtime computing*" (Hansen, 2004, emphasis in the original). Raymond Williams expressed the same sentiment a generation before:

> However dominant a social system may be, the very meaning of its domination involves a limitation of selection of the activities it covers, so that by definition it cannot exhaust all social experience, which therefore always potentially contains space for alternative acts and alternative intentions which are not yet articulated as a social institution or even project.
>
> (Williams, 1979, p. 252)

Space and time as social processes are transformed through their compression by globalization, but this globalization is deeply political and so, once again, any countering of its effects must also be political. Temporality as the hidden dimension of political sovereignty holds the key to the political impasse, and it is to the concept of time sovereignty that we now turn.

Temporal sovereignty

The idea of spatial sovereignty is well understood. It has a long history in political theory and political philosophy, and it functions still as a fundamental category of democratic theory. The concept of temporal sovereignty is not so easily accepted or understood. Why?

In his historiography of the concept, Jens Bartelson's *A Genealogy of Sovereignty* (1995) gives a compelling explanation for our lack of appreciation for the *temporal in the political*. Through meticulous research, Bartelson show that the eliding of time from the idea of sovereignty was a deliberate political choice. He writes that time is one of the "unthought foundations of our political knowledge" when related "to the concept of sovereignty" (Bartelson, 1995, p. 4). He shows that time was made to *disappear* from sovereignty's praxis during the phase of its early-modern inception. Time is neither implicit nor explicit in the question of sovereignty, but instead has become a taken-for-granted backdrop to it. Bartelson argues that time as understood in late antiquity was a thought-world dominated by Augustine, whose "solution" to its difficulty for human comprehension was to relegate it from the realm of the cosmological to the subjective and ephemeral human world (Bartelson, 1995, p. 96). However, this had problems for the legitimacy of sovereign power and authority where it was undermined by the "stigma of transitoriness" (Bartelson, 1995, p. 101). If people can become weak and die, then the state and its authority can also, is how the logic ran. However, political theory in the time of the Renaissance looked back to Aristotelian conceptions of cyclical and recurrent time in order to give the political world seeming permanence. Time based upon celestial movements, on the rhythms of seasons, and increasingly upon the clock, served to abstract humans from time's experience. People instead began to imagine that they *exist within* this unending time. Correspondingly, the sovereign body politic, or the person of the sovereign ruler, through their distance from ordinary people and life, "could be seen as something continuous in time" (Bartelson, 1995, p. 98). Abstract and taken-for-granted time in which ordinary humans are not implicated is thus able to convey *permanence and authority* upon the body politic to the point where, as Bartelson notes, "it lives forever" (Bartelson, 1995, p. 99). In short, for time to be seen as something abstract and infinite and associated with the body politic, is in practice to abstract social power from the fragility and createdness of earthly existence. In terms of the function of power, elites could not allow time to be viewed as a sovereign realm. To do so would be to render time *social*, a site of contestation, of possible transformation and potential democratization.

Bartelson's *Genealogy* is a conceptual breakthrough whose value is only becoming apparent in our age of network time, and it gives new perspective to the classic conundrums of political sovereignty hitherto understood in terms of space only. In the immediate post-1945 period Hannah Arendt, in her book *The Origins of Totalitarianism*, identified a central contradiction flowing from the theory and practice of lop-sided space sovereignty. In a context where refugees and displaced persons were a significant problem due to the war's disruption, Arendt argued that the theory of political sovereignty as a universal human right was corrupted by the nation-state acting to normalize and institutionalize the process of "rightlessness" and the "loss of political status" for refugees (Arendt, 2004, p. 374). The universal ideal of human rights whereupon everyone regardless of nationality is guaranteed the protections of sovereignty, is routinely trumped by nation-state power that prioritizes perceived national interests (and the rights of its "own" people) over stateless persons. The state distinguishes, in other words, between national and non-national, with legal protection extending only to the former. Non-nationals languish in transit camps, or endure legal limbo, sometimes for decades, because nation-states are reluctant to see the stateless as their problem, as people whose rights should be automatically guaranteed. The nation-state versus universal human rights contradiction, Arendt theorizes, is not an aberrance provoked by post-war disorder, but a *structural contradiction* that marks a fatal flaw in the logic of sovereignty. She writes:

> No paradox of contemporary politics is filled with more poignant irony than the discrepancy between the efforts of well-meaning idealists who … insist upon regarding as "inalienable" those human rights, which are enjoyed only by citizens of the most prosperous and civilized countries, and the situation of the rightless themselves.
>
> (Arendt, 2004, p. 355)

That the problem of statelessness and rightlessness persists and worsens today is powerful testimony to this "paradox." Indeed, and as I touched on previously, the processes of globalization, far from undermining the power of the nation-state vis-à-vis its own people, can actually strengthen it to the point where it routinely turns upon its own to undermine localized sovereign rights and protections such as privacy, freedom of speech, provision of affordable education, decently paying jobs, a functioning civil society, and so on. The state in "the most prosperous and civilized countries" has not abrogated political power in these areas, but intervenes, through anti-democratic legislation of all kinds, to shape and direct the lives of millions for whom neoliberalism delivers least and yet affects most.

To consider time as sovereign would enable us to review the paradox that Arendt identified. The new context for this appears through the identification of our new relationship with time, network time – and to see this itself as being a paradox. As we saw, in the modern period, space and sovereignty were based upon the idea of the territorially defined nation-state, but the key concept of *universality* is structurally delimited here. However, if we recognize time as created, as social and as rooted in the experience of people in the world, then its new and powerful network form represents a *universality of the experience of lived time* in ways that the zoned and predictable clock time world of modernity could not. The paradox is that to view the experience of time as a universal experience through globalization makes time salient and political, but it also reveals it as *unsustainable.* I will return to this point presently.

As noted before, acceleration is the central dynamic of network time, and this is driven by competition at the local and global level, creating a 24/7 form of capitalist accumulation that spans the entire planet (Crary, 2014, pp. 29–60). The existence of network time in this neoliberal form is universal in that almost everyone feels the effects of its acceleratory logic. Much work has been done already in time studies that consider the effects of what Dale Southerton terms the "time-squeeze" (Southerton, 2003), emanating from the logic that Ben Agger (1989) describes as acting like a "time bandit," to create what Juliet Schor sees as an omnipresent form of human oppression in her landmark *The Overworked American* (1998). The "consumer treadmill" (Schor, 1998, p. 38) processes she describes, of course, extend beyond the US context to affect the entire neoliberal sphere. What is also clear is that the time oppression wrought by the networked economy is cross-class, cross-cultural as well as cross-border, and in that sense is universal both at the technological and human levels (Hassan, 2012). This *universal temporal domination* is a largely unrecognized political question. However, if we consider it alongside Arendt's paradox, we can see that it too constitutes a form of rightlessness, where sovereign autonomy over lived time is expropriated and commodified through its digitalization to an intensity and extensity not possible under the modern regime of the clock. Its universal aspect through the networking of time is what brings it to the light – to where we can begin to see it as the basis for the construction of a properly balanced sovereignty that encompasses both space and time.

Toward a theory of temporal sovereignty for a global age

It may seem that the barriers to thinking in terms of sovereign time are too ancient and insuperable for the concept to have political purchase, even as a nascent idea. However, network time's universality, coupled to its economic and social unsustainability, presents footholds from which we may progress. In terms of the unsustainability paradox, the footing is firm. Network time as lived creation and experience of time is bound tightly to the logic of capitalism, Stiegler's "commodity time." This is unsustainable in precisely the way that neoliberal capitalism is. That is to say, it constitutes the enormous waste of resources on an unprecedented scale, and (following from this) is destructive of the environmental commonwealth and ecological well-being of the planet. Such lived time cannot be seen as positively oriented time. Unsustainability looms at the human level also. The acceleration of network time is untenable because humans have speed limits and computers do not. When computers are networked and driven faster by the imperatives of profit, humans become the weak link in this human-computer interaction. The rhetoric of "multitasking," for example, has been exploded by neuroscience, which shows that we cannot do it well (Carr, 2010, pp. 140–141). Moreover, the gains to our cognitive functioning in a rising sea of information have been identified as relatively modest (Carr, 2010, p. 133). In these and many other areas of neoliberal life, the quest for economic "efficiency" through faster and faster ways of producing and consuming information, creates diminishing returns from humans whose capacity to function as political beings and as members of a civil society has a low threshold when compared to networked computers.

The durability of both economic systems and human well-being are prime reasons to be vitally interested in viewing time as a realm of sovereignty, as an inalienable element of what it is to be fully human – and therefore something in need of political recognition and protection as a human right. Without such, the best we could hope for is to ride the boom and bust cycles of capitalism, and let its disorder determine the shape, direction (and speed) of our lives. Political action as the means of redress of rights needs to be focused upon the temporal as well as the economic and social. Positive political action comes only through recognition of a problem, an injustice, and an opportunity. Under neoliberal capitalism, network time is all of these. Some of the problems and injustices of social and economic acceleration are recognized already at the multilateral and local levels – and real opportunities may arise if sufficient *political will* can flow from sustained and reflective thinking about the function of time in our globalized post-modernity.

We can only deal with a couple of examples here. First, the widely perceived problem of a damagingly accelerated stock market system that almost brought down the global economy in 2007 could be ameliorated at the structural level. The so-called Tobin Tax, or tax on financial transactions, if instituted would not only slow down and make more reasoned and responsible the frenzied trade that underpins globalization, but would also be a source of considerable revenue at the multilateral level and be available for positive uses (Traynor, 2011). Second, at the everyday level, the problem (and injustice) of so-called work-life balance has simmered in policy circles, in media and in business for some time. The problem is that there is no consensus regarding the definition and framing of this late-capitalist phenomenon. The process itself is well understood: work has colonized the sphere of life, of family, of evenings and weekends, and of leisure more broadly, to an unprecedented degree. However, the debates tend to center on the vacuous idea of "balance." What it is that needs to be in balance – the subjective time of the individual – is never made problematic in terms of time's *ownership* and time's fundamentally *political* nature. Even so, businesses as well as workers view this at least as a social and economic issue. Facebook executive Sheryl Sandberg, for example, agrees there's a problem, and advocates for more "balance," but does not specify what it is that needs balancing. She opined: "there's no such thing as work-life balance. There's work, and there's life, and there's no balance" (Bosker, 2012). There is no mention of time, either. If we do make time salient then the nature of the imbalance becomes clear and is seen to emanate from the demands of globalizing capital and its imperatives of "flexible accumulation." Clearer still, the lack of balance is due to what Harvey terms "accumulation through dispossession" – which in the network society is the systemic deprivation of the subjective time of the individual (Harvey, 2005, pp. 137–183). Harvey is at pains, through a reading of Rosa Luxemburg and Hannah Arendt, to show that such "dispossession" has been often occasioned by brute force, both historically and in our own time. This means, then, that work-life balance problems must also be seen in their class context: accumulation by dispossession is relative, and dispossession of time's ownership by those Sheryl Sandberg has in mind differ in kind and degree from, say, the low-paid worker who works 60 hours per week and still cannot make ends meet.

These are only two examples, but they go to the heart of the crisis of time that accompanies the crisis of space in our globalized age. Martin Heidegger (1972, p. 16) wrote that "there is no time without man," meaning that we cannot understand time without recognizing its human-subjective essence. We can recognize time as economic, and we also can understand it as being social, but it is necessary to be rid of the elision between time and sovereignty that Bartelson identifies, and see time as political too – perhaps as *fundamentally* political. This would achieve a balance between space and time and in how we understand and practice the concept of sovereignty. Given the fact that these concepts are so deeply linked to the history and trajectory of capitalism, to politicize time and see it as human and as residing in the ontological core of everyone, would be a step towards not only humanizing globalization, but also humanizing capitalism. The madness of open-ended acceleration of economy and society, where competition leads only to unending demands for more speed, could be seen for what it is: irrational at the level of economics, creating an increasingly volatile and remote-controlled global economy that may fail at any time; and irrational at the level of the individual who is necessarily disempowered by networked systems that are developed with the abstractions of profit and efficiency and the substitution of human brain and labor power written into their very code (Ashworth, 1996).

Temporal sovereignty as a legitimate human need and right in and of itself would be no panacea to the problems of post-modern globalization. The imbalance between state-based sovereignty and the universal ideal would not be resolved easily. The crisis of space that is at the core of capitalism's expansionary logic will remain as long as capitalism functions as a colonizing force – colonizing of space and colonizing of human relations that are not democratically robust enough to resist. However, the computer logic of globalization itself gives hope here. The *virtual* space that is created by networks is limitless. David Harvey argued in his *Limits to Capital* that what he termed the "spatial fix" has acted as a kind of provisional safety valve against the pressures of over-accumulation and historically enabled capital to survive in its classical mode until the 1970s (Harvey, 1983, pp. 411–451). Virtual space could function as a permanent spatial *solution* – in the context of a more humanized form of capitalism and globalization. Capital surplus needs to be invested somewhere, and cyberspace has provided the unexpected means for the needed spatial displacement – both to be realized as profit, and in theory to orient the network toward more social ends. What I have elsewhere termed a "temporalized internet" (Hassan, 2014) could function to make the network more responsive to both commercial and social needs in ways that are more human-centered, using the spatial-temporal potential of a truly "social network" oriented toward social-democratic ends.

Some conclusions

The merging and balancing of the "unhelpful dualism" that constitutes thinking on space and time, as economic, social and political concepts must be the initial conceptual step. A skeptic regarding the utility and future of sovereignty, Hannah Arendt can nonetheless help us take a step toward recognition of time as sovereign. Writing about Arendt's thinking on political rights, Judith Butler observes:

> In … "The Decline of the Nation-State and the Rights of Man" Arendt outlines what she takes to be the essential preconditions for the exercise of any rights at all. And these preconditions include *place* and *political belonging*. She writes: "the fundamental deprivation of human rights is manifested first and above all in the deprivation of a place in the world which makes opinions significant and actions effective."
>
> (Butler, 2014, p. 147, emphasis in the original)

This is valuable. It is more so if we balance Arendt's "place" with a view of the *lack of time as a category of deprivation*. Human time, as well as human place, must become a precondition for the exercise of political rights – and for a richer sense of political belonging – or these can never be fully attained.

With network time we have moved beyond the unreflected-upon clock as backdrop to modern life, to a dynamic and volatile and accelerating time that colonizes and harries our lived time, the time that is subjective and fragile and has limits. Network time is largely negative time, notwithstanding the claims made for the network per se by the multinational purveyors of computer systems and digital applications. The opportunity that arises is that the idea of time and its colonization by capitalism is becoming significant as a post-modern phenomenon. We need to see in its new context a process that is human-centered instead of technologically dominated, and is therefore political. If we do think of this as a political question of unfairness, of unsustainability and of time deprivation, then it is a short conceptual step to think about issues of rights and sovereignty.

In a recent essay on Arendt and political sovereignty, Patrick Hayden emphasizes the "awkward coupling" of individual and nation-state conceptions of sovereignty (Hayden, 2007, p. 251). Through Bartelson I have tried to show that this is still a problem because time was not balanced with space in the ancient conception of sovereignty. Consequently time became the hidden aspect of political sovereignty. The suppression of political time was something that was manageable, especially through the domination of a naturalized clock time, until only very recently. The context for the recognition of the imbalance between space and time emerges with the rise of neoliberal globalization and the information technology revolution, but the compression of space and time through globalization becomes a political opportunity only if we review the nature of the space-time transformation, and see that both dimensions are fully human-centered. To put it simply: space, if we follow Lefebvre (1991), is social before it is physical; and time is human before it is technological. Both are thus political, and both warrant protection as a singular sphere of human rights – as a sovereign realm.

Where to begin? Arendt can help us one more time. Following Rousseau, she argued that "[f]reedom as an inner capacity of man is identical with the capacity to begin" (Arendt, 2004, p. 473). We are born free but become enslaved. However, the capacity to begin in some form is always there and is like the capacity to be "born" again – and so to begin again. Slavoj Žižek considers a similar point, arguing that in order to rescue ourselves from the present monopolitical age, we need to go back to our conceptual beginnings, to take stock, and then to choose a different path from which we can move forward once more (Žižek, 2009, p. 51). The very staleness of what passes for liberal democracy today must force a wholesale reconsideration of the nature of its politics. To go back to the very beginnings of political theory in Europe, and the conscious political decision to elide time from considerations of what constitutes the sovereign, would place Arendt's paradox in a new light. The acceptance of the sovereignty of time would add a new dimension to our understandings of what it is to be human and political and to be living on the face of the physical Earth as well as within its digital-virtual correlate. A new beginning and a new path would thus constitute the basis for fresh political agendas; for a political action on immediate and longer-term space-time problems – which are many and varied. The unique thing about a rebalancing of space and time in the question of sovereign rights is that it is – thanks to networked globalization – a powerfully *universal* concern, and it is the emphasis upon this point – in media, in forums, in political action of every kind, online and off, from the workplace to the home and from civil society to the institutions of politics – that can galvanize the billions for whom temporal autonomy – or even simple "free time" diminishes day by day. Neoliberal globalization atomizes individuals and homogenizes political thinking, but globalization also provides the opportunity to rethink the question of sovereignty with the added dimension of time. Political blueprints are inherently authoritarian, but to expand the scope of political possibility is not. So in more ways than we can know today, a temporalization of politics of the sort described here can bring people to consider the fuller question of rights, and so to diversify their thinking on how best to promote these through democratic processes in space and time.

Note

1 Of course the concomitant to Linklater's rather breezy and Western-centric analysis was 400 years of bloody colonialism and its bloody reversals in the 20th century. Linklater, like many more conservative historians, makes more of the "creative" processes in this context as opposed to the "destructive." However, as Eric Wolf relates in his masterly *Europe and the People Without History*, political sovereignty was indeed for export, but it was primarily the "extension of European sovereignty abroad" (Wolf, 1997, p. 313).

References

Adam, B. (2004) *Time*. Cambridge: Polity.

Agger, B. (1989) *Speeding up Fast Capitalism*. New York: Paradigm Publishers.

Anderson, P. (2000) Renewals. *New Left Review*, 1, pp. 5–24.

Arendt, H. (2004) *The Origins of Totalitarianism*. New York: Schocken Books.

Ashworth, W.J. (1996) Memory, Efficiency and Symbolic Analysis: Charles Babbage, John Herschel, and the Industrial Mind. *Isis*, 87(4).

Bartelson, J. (1995) *A Genealogy of Sovereignty*. Cambridge: Cambridge University Press.

Bauman, Z. (2000) *Liquid Modernity*. Cambridge: Polity.

Bosker, B. (2012) There's No Such Thing as Work-Life Balance. *Huffington Post*, December 4. www.huffingtonpost.com/2012/04/06/sheryl-sandberg_n_1409061.html.

Brown, W. (2010) *Walled States, Waning Sovereignty*. New York: Zone Books.

Butler, J. (2014) *Parting Ways*. New York: Columbia University Press.

Carr, N. (2010) *The Shallows: How the Internet is Changing the Way We Think, Read and Remember*. London: Atlantic Books.

Cohan, W. (2011) *Money and Power*. New York: Doubleday.

Crary, J. (2014) *24/7*. London: Verso.

Fukuyama, Francis (1989) The End of History? *National Interest*, 16 (Summer), pp. 3–18.

Gleick, J. (2000) *Faster: The Acceleration of Just About Everything*. New York: Abacus.

Hansen, M. (2004) Realtime Synthesis. *Culture Machine*, 6, www.culturemachine.net/index.php/cm/article/view/9/8.

Harding, L. (2014) *The Snowden Files*. New York: Vintage.

Harvey, D. (1983) *The Limits to Capital*. London: Verso.

Harvey, D. (1989) *The Condition of Postmodernity*. Cambridge: Blackwell.

Harvey, D. (2005) *The New Imperialism*. London: Oxford University Press.

Hassan, R. (2003) Network Time and the New Knowledge Epoch. *Time & Society*, 12(2–3).

Hassan, R. (2009) *Empires of Speed*. Leiden: Brill.

Hassan, R. (2012) *Age of Distraction*. New York: Transaction Books.

Hassan, R. (2014) A Temporalized Internet. *Political Economy of Communication*, 2(1), pp. 3–16.

Hayden, P. (2007) From Exclusion to Containment: Arendt, Sovereign Power, And Statelessness. *Societies Without Borders*, 3, pp. 248–269.

Heggestuen, J. (2013) One in Every Five People Owns a Smartphone. *Business Insider*. www.businessinsider.com.au/smartphone-and-tablet-penetration-2013-10.

Heidegger, M. (1972) *On Time and Being*. trans. Joan Stambaugh. New York: Harper.

Himmelfarb, G. (2012) Civil Society Reconsidered. *The Weekly Standard*, 17(3). www.weeklystandard.com/articles/civil-society-reconsidered_636994.html.

Hobbes, T. (2011) *Leviathan*. New York: Broadview Press.

Hutchings, K. (2008) *Time and World Politics*. Manchester: Manchester University Press.

Judt, T. (2005) *Postwar*. London: Vintage.

Klinke, I. (2013) Chronopolitics: A Conceptual Matrix. *Progress in Human Geography*, 37(5), pp. 673–690.

Kothari, M. (2005) Homelessness Increasing All Over the World. *Washington Times*. www.washingtontimes.com/news/2005/apr/10/20050410-105739-5991r/?page=all.

Lefebvre, H. (1991) *The Production of Space*. Oxford: Blackwell.

Linklater, A. (2014) *Owning the Earth*. London: Bloomsbury.

Marcuse, H. (1990) *One-Dimensional Man*. Boston, MA: Beacon Press.

Milne, S. (2004) *The Enemy Within*. London: Verso.

Ó Tuathail, G., Dalby, S. and Routledge, P. (2006) *A Geopolitics Reader*, 2nd edn. London and New York: Routledge.

Rosa, H. (2013) *Social Acceleration: A New Theory of Modernity*. New York: Columbia University Press.

Scheuerman, W. (2004) *Liberal Democracy and the Social Acceleration of Time*. Baltimore, MD: Johns Hopkins University Press.

Schor, J.B. (1998) *The Overworked American*. New York: Basic Books.

Southerton, D. (2003) Squeezing Time. *Time & Society*, 12(1), pp. 5–25.

Stiegler, B. (1998) The Time of Cinema. *Teknema: Journal of Philosophy and Technology*, 4.

Stiegler, B. (2011) *For a New Critique of Political Economy*. Cambridge: Polity.

Stiglitz, J. (2008) The End of Neoliberalism? *Project Syndicate*. www.project-syndicate.org/commentary/the-end-of-neo-liberalism?barrier=true.

Thrift, N. and May, J. (2002) *Timespace*. London: Routledge.

Tomlinson, J. (2007) *The Culture of Speed*. London: Sage.

Traynor, I. (2011). EU calls for Tobin Tax. *The Guardian*, June 29. www.theguardian.com/world/2011/jun/29/ec-proposes-tobin-style-taxes.

Vaughan-Williams, N. (2009) *Border Politics: The Limits of Sovereign Power*. Edinburgh: Edinburgh University Press.

Virilio, P. (1995) Speed and information: Cyberspace Alarm!, www.ctheory.net/text_file.asp?pick=72.

Wallis, G. (1970) Chronopolitics: The Impact of Time Perspectives on the Dynamics of Social Change. *Social Forces*, 49(1), pp. 102–108.

Williams, R. (1979) *Politics and Letters*. London: Verso Books.
Wolf, E. (1997) *Europe and the People Without History*. Berkeley: University of California Press.
Žižek, S. (2009) How to Begin from the Beginning. *New Left Review*, 57, pp. 43–55.

3 Time, worker exploitation, and global capitalism

Wayne Hope

Among those who advance globalization as a worthwhile subject of analysis, three positions suggest themselves. The first affirms that human migration and trade routes have long been global in scope; one can map the socio-geographies of globalization against the *longue durée* of history. Second, it can be argued, following world systems theory, that globalization originates with the geographic expansion of market capitalism during the 16th century and that current world networks are not substantially new. The third position, which informs this chapter, foregrounds the epochal distinctiveness of global capitalism[1] and its associated networks of worker exploitation. More than a geo-spatial explication is required here. In line with the theme of this volume, I will examine the time-related aspects of worker exploitation in a globally networked world. The salient categories of analysis are time reckoning, temporality, and coevalness. These categories will underpin a critique of labor and labor time arbitrage, supply chain subcontracting and precarity, plus the financialization of corporate activity in regard to employer-worker relationships. Together, these global capitalist processes constitute a time-based schema of worker exploitation. After sketching the strengths and weak points of this schema, I will consider recent manifestations of global worker activism from a time-related perspective.

Global capitalism and the global worker

Global capitalism is an epochal phenomenon with historical antecedents. The early development of mass communication, telegraph, and telecommunication networks from about 1860 provided the infrastructure not just for Western empires and national economies but for world capitalist expansion (Winseck and Pike, 2007). After 1945, with decolonization, capitalism expanded into newly independent countries amidst a Cold War setting. Global capitalism per se began to take shape in the 1990s after the disintegration of the Bretton Woods currency system, national Keynesianism, the Soviet Bloc and Third World liberation movements. The associated international expansion of neoliberal policy regimes propagated the assumption that human well-being could best be advanced by deregulating national economies, privatizing state assets, and commercializing public institutions.

According to political economist William Robinson (2004), such developments encouraged the growth of transnational corporations (firms with headquarters in more than three countries). His analysis draws from various world investment reports published by the United Nations Conference on Trade and Development (UNCTAD) and from privately commissioned financial reports outlining the growth of such corporations alongside the increase in cross-border mergers/acquisitions. The latter trend suggests that for the largest corporations, the distinction between home and host country has weakened. Between 2000 and 2007 inclusive, cross-border merger and acquisition deals in excess of US$1 billion totaled 1,335 (compared to 479 from 1992 to 1999) (UNCTAD, 2008).

In general terms the globalization of capitalism allowed major investors and transnational corporate executives to speed up profit rates, cross-tabulate profit ratios and contract the chronological parameters of profit measurement. The entire process began with the mutual expansion of finance capital and information and communication technologies (ICT) infrastructures. Once government control over exchange rates and capital movements became unsustainable, during the 1970s, private financial institutions were able independently to generate credit and currency units. The subsequent globalization of finance capital depended upon microchip computers and associated advances in telecommunications. Satellite, optical fiber and cellular networks expanded and accelerated information transfer between computer terminals. All of these developments created an unprecedented economic environment; multiple currency units and complex financial assets worth trillions of dollars were traded globally in real time (Castells, 1996, 2000, pp. 52–75; Held et al., 1999; Singh, 2000). The proliferation of ICT infrastructures also globalized networks of management, production and distribution. Corporates allow affiliates greater autonomy and encourage channels of two-way communication (e.g. via email, e-commerce, intranets, video conferencing, mobile telephony and inter-enterprise computer networking). Transnational corporations use information and communication systems to decentralize and subcontract operations and to centralize financial control. This pattern is common to otherwise separate sectors of production, for example textiles, garments, automobiles and semi-conductors (Dicken, 2004, pp. 317–436). At the same time, major firms in the same industry cooperate through joint ventures, subcontracting, franchising and research projects that will design specific products for identifiable markets. For Manuel Castells, these innovations were indicators of the "network enterprise." Within this organizational model, firms remain the primary unit of capital accumulation, property rights and strategic management while routine business practices are performed by flexible ad hoc networks (Castells, 1996, 2001, pp. 67–78).

The globalization of finance capital and the proliferation of transnational "network enterprises" were interrelated developments. The process dates from the 1980s when new kinds of financial market activity reshaped the American corporation. Hostile takeover firms grew in number and activity. They broke up conglomerates so that component parts could be sold for more than the previous market valuation. Individual investors gave way to institutional investors who gained major control over corporate stock. Such investors did not simply value conglomerates as a whole; priority was given to divisional profit performance and the establishment of "focused" firms. Consequently, share analysts grew in number and developed specializations. Corporations eager to attract favorable analyst coverage were obliged to sell off businesses unrelated to core competencies (Zorn and Dobbin, 2005). The core competency shareholder value priorities of institutional investors and analysts began to shape corporate governance and strategy. The chief executive worked with the chief financial officer to generate profits from core competencies and to manage the share price, in accordance with the preferences of institutional investors and analysts. In these circumstances priority was accorded to the profit and share price estimates agreed upon by analysts. A corporation's performance thus depended on whether it could meet or beat the estimates. At the same time, specialist teams of analysts frequently announced buy and sell recommendations. The net result was share price volatility and a short-termist culture of corporate governance. As neoliberal policies gained international credence, so too did the financialization of corporate governance. By the early 2000s, this short-termist mode of governance was beginning to confront, and intermesh with, the longer-term strategic priorities of Japanese, South Korean and Taiwanese corporations along with Chinese state enterprises. They worked to sustain stable inter-firm and state-corporate relationships based upon shared obligations (Hope, 2009, pp. 69–71; Harris, 2012, pp. 220–231).

Global capitalism is more than a collection of national economies and production systems related through external exchange. Transnational corporations have developed worldwide commodity chains based on subcontracting, outsourcing and allied arrangements. Anna Tsing has described this phenomenon as "supply chain capitalism" (Tsing, 2009). The term may be an overgeneralization. As I will later explain, supply chain capitalism is a phenomenon that connects with financialized capitalism and informational capitalism (as encapsulated by the network enterprise). What cannot be doubted, however, is that supply chains are pervasive. Here, Tsing provides a compelling depiction:

> Supply chains offer some of the most vivid images of our times: telephone operators assisting customers from across the globe; "traditional" indigenous farmers growing specialist crops for wealthy metropolitan consumers; Chinese millionaires reaping the profits of Walmart contracts; sweatshop workers toiling in locked rooms while brand name buyers disavow responsibility.
>
> (Tsing, 2009, p. 149)

These images direct our attention to what Nick Dyer-Witheford calls the "global worker." This figuration, in contradistinction to the Fordist blue-collar worker from the Western industrial heartlands, is central to the globalization of capitalism. The global worker is variegated by an increasingly complex division of labor strongly associated with the service sector, universalized by the incorporation of women workers, the growth of production centers outside the West and by flows of migrant labor. New jobs and occupations, with many hierarchies, are connected directly to ICTs as indicated by the rapid growth of Internet and cell phone use. The computer industry, for example, contains a software sector incorporating business applications and digital games designed and engineered in North America, Western Europe and Japan. Programming jobs have been typically outsourced to subcontractors in Eastern Europe, South Asia and Southeast Asia. In the hardware sector, salaried engineers and architects design and prototype phone, gaming and specialist computer devices. Assembly of these devices has been performed in Central America, Eastern Europe and southern China. Manual labor also entails mining the specialist minerals necessary for consumer electronics and excavating parts from toxic e-waste disposal sites in Asia and Africa (Dyer-Witheford, 2010, 2015; Roth, 2010; Fuchs, 2013).

The variegated social lifeworlds of the global worker are shaped by the forces of global capitalism and regional/local traditions including the cultural legacies of Western colonialism. To understand how these forces play out in particular circumstances we need a conception of exploitation that acknowledges the pluralities of time. It is to this task that I now turn.

Workers, exploitation and epistemes of time

Let us now consider the concept of exploitation. Economically, this occurs when one section of the population controls, and benefits from, the labor of another section. In conventional Marxist terms exploitation occurs when the class of industrial capitalists, within a capitalist economic system, generates profit from working-class labor. In this situation workers have only their labor to sell. Other exploiting classes or class fractions benefit from this arrangement. Erik Olin Wright provides a more analytical description of exploitation. In his view, the material welfare of the capitalist class causally depends upon the deprivation of the working class. Exploited workers are excluded from access to productive resources and private assets by the legal enforcement of property rights. Capitalism does not simply deprive and marginalize workers. It also enables capitalist owners of productive resources to appropriate their labor. The efforts of the exploited ensure the welfare of the exploiter (Wright, 1997, pp. 10–11).

The component features of exploitation explain the inseparable relation between capitalist profits and workers' wages. This relation materially unfolds as a class contest over rights and remuneration. The social realities of exploitation, I want to suggest, can also be characterized as "a bitter and dark struggle around time and the use of time" (Lefebvre, 2004, p. 74). This observation from French philosopher Henri Lefebvre was intended to encapsulate a vital truth about everyday life in general. It contains conflictual modalities of time associated with bodily rhythms, solar rhythms, biological nature, biospheric seasonality, and social custom. In the modern world, these modalities of time experience are in constant struggle with the obligations and schedules associated with quantified sequential time. Such a struggle is inherent to, and formalized by, the processes of class exploitation within capitalism. The darkness and bitterness of it all is mainly experienced by workers; capitalist employers have more time resources at their disposal. Yet, the capitalist can never disengage from the struggle. The appropriation of labor and the generation of profit, as measured by time, cannot be guaranteed. Here, one must acknowledge that time is not a singular construct. Rather, there are different epistemes of time and each of them illuminates certain aspects of worker exploitation in a capitalist economy. Earlier, I drew upon the episteme of epochality to identify the historical distinctiveness of global capitalism and the global worker. Against this background, I will now consider the epistemes of time reckoning, temporality, and coevalness in relation to exploitation.

Time reckoning

This episteme refers to the naming and numbering of time independently of intimate, socially lived time, and of inbuilt natural processes within geology, biology, the biosphere, and the solar system. Of course, the reckoning of time cannot be entirely divorced from social or natural settings. Calendar time, for example, is closely associated with the diurnal cycle and seasonal changes of climate. The level of abstraction in time reckoning indicates the extent to which the relevant measures overrule the macro-processes of nature. The named and numbered units of clock time for example, constitute a system of time reckoning which incorporates or marginalizes other appreciations of time. In this respect, naming and numbering entails the objective management of time via the process of ordering, structuring, division, classification, synchronicity, and prediction. Subdividing time according to years, months, weeks, days, hours, minutes and seconds invokes ordered sequentiality. This sense of time became variously symbolized and culturally embedded by the instruments and devices of time keeping: sundials, calendars, mechanical clocks, and timepieces. By the mid-19th century, these innovations had become the reference points for commercial enterprises, factories, offices, trading networks, and early-modern state institutions. Increasingly standardized measures of sequential time defined the routines of public and private life at the expense of seasonal and biological time. The standard demarcations of clock time were exported worldwide through the establishment of railway and steamship networks. In 1884 the inaugural International Meridian Conference agreed to standardize time zones across the globe (Adam, 2004, pp. 111–117).

Critically, this ascendant regime of time reckoning confronted the arrivals of telegraphy, telephony, and wireless broadcasting. With these innovations, a new kind of time reckoning emerged; speed of transmission across networks transcended the durational and sequential properties of clock time. Subsequent technological advances such as satellite television, inter-networked computers, the Internet, mobile telephony, and social media would usher in a new regime of time standardization. Within this regime, speed cannot be measured by the time taken to cover geographic distance. Electronic speed transcends the obstacles and frictions of physical geography. Across networks, this kind of speed is measured against the standard of instantaneity. Thus, efficient transmission is not assessed within the parameters of sequential clock time. Instead, the central question is – can lengths or lags of time be reduced or eliminated? This drive toward instantaneity, within and between networks, is encapsulated by the notion of real time. Worldwide proliferations of real time work to disrupt, usurp, or colonize clock time, processes which engender the widespread feeling that life is accelerating.

For capitalist enterprises, calculations of profit and labor productivity came to depend upon the standard demarcations of clock time. The process is delineated in E.P. Thompson's seminal article, "Time, Work Discipline and Industrial Capitalism" (1967). In his account, clock time was a nascent presence within Medieval and mercantile Europe before converging with the growth of industrial capitalism. In non-industrial economies centered around hunting, farming, fishing and trade, work is task oriented and embedded within the social fabric of family, tribe, village and town. As Thompson points out, "social intercourse and labour are intermingled. The working day lengthens or contracts according to the task, and there is no great sense of conflict between labour and 'passing the time of day'" (Thompson, 1967, p. 60). The social lifeworlds of task orientation are instilled with the rhythms of nature (weather, seasons, the diurnal cycle) and biology (circadian, hormonal and reproductive cycles). By contrast, timed labor requires the synchronization of work tasks. This emerged in 18th-century England as small, domestically based workshops developed simple divisions of labor. The eventual arrival of machine-powered industry in the mid-19th century required greater exactitude in time routines and a more substantial synchronization of labor processes. Waged labor and time discipline was most rigorously imposed in textile mills and engineering workshops. Thompson notes that this was an attritional process: "the first generation of factory workers were taught by their masters the importance of time; a second generation formed their short-time committees in the ten hour movement; the third generation struck for overtime or time and a half." By this stage, organized workers had "accepted the categories of their employers and learned to fight back within them" (Thompson, 1967, p. 86). An associated propaganda campaign of time thrift was advanced by employers, Church leaders, school teachers and conservative politicians. Their insistent message was that time must be put to use; it was offensive for men and women of working age to merely "pass the time" (Thompson, 1967, p. 61).

From the early 20th century, clock-time measures of work and production became a systematic exercise. Frederic Taylor's doctrine of scientific management was designed to eliminate time waste on the production line and to exclude workers from the development of cost-saving production systems. This pattern of exploitation was legitimized by the organizational cultures of management and the academic discourses of management studies (as encapsulated by the seemingly anodyne requirements of "productivity"). In given circumstances, the excesses of Taylorism could be challenged by legally constituted union organizations acting on behalf of their members across workplaces, occupations and industries. At national level, union federations could be part of tripartite negotiations over wage determination with government leaders and peak employer groups.

Within the ICT networks of global capitalism the scale, speed and precision of Taylorist work flow calculations are unprecedented. Factory and office managers of the industrial-Fordist era were incapacitated by comparison with their contemporary counterparts. Most importantly, the ascendancy of ICT-enabled real time over clock time has fundamentally altered the core dichotomy between capital and labor. Global capital operates in real time as workers continue to labor in clock time (Hoogvelt, 2010, p. 54). Transnational corporations can decentralize operations while centralizing control over workforces. Corporate executives can peruse on-screen spreadsheets displaying the performances of factory machine operators in different workplaces. Networks of real-time communication and information retrieval allow senior managers to identify and remove lags of sequential clock time from their production systems and subcontracted franchises. Thus, it is difficult for dispersed global workers to employ the categories of clock time to improve wages and conditions against remote corporate executives and managers. Within the garment industry, for example, factory workers in each exporting country confront local alliances of subcontractors, plant owners, politicians, police and security firms. They enforce transnationally calibrated employment regimes set by the world's largest apparel brands. Internationally competing pools of garment workers exert downward pressure on apparel sector wages in most countries (of late, the exceptions to this trend are China, India, Indonesia and Vietnam) (Goria, 2014, pp. 29–33, 35–61, 366–386).

Temporality

Temporality is not about the dissecting, dividing, counting, or measuring of time. Rather, it is the ground for thinking about the relationalities which connect past, present, and future. Temporality thus incorporates memory, expectation, and attention to the present. These are manifestations of lived time that have no necessary connection to any supervening conception of time deriving from religion or science (Munn, 1992, p. 94). Temporality can be observed or experienced in the interacting domains of self, intersubjectivity, affective identity, institutions, and society at large. Temporal understandings are also intrinsic to the routine operations of nation-states, economies, legal systems, and polities.

Further examination reveals fundamental differences of temporal orientation. Past-oriented temporality recognizes that the past could not be constituted if it did not coexist with the present (whose past it is). From this perspective, the past, through which all presents pass, can only exist in the present. The indivisibility of past and present activates the process of remembering, although the content of remembrance may be contested from different points of view over time. By contrast, future-oriented temporality emphasizes the mutually interacting relationalities between present and future. The results of such interactions are manifold. Studies of the future from the standpoint of the present incorporate futurism, futurology, and science fiction. Yet, futures also live in the present as is encapsulated by the phrase "futures in the making." Points of reference here include economic, technological, and architectural projects that entail originality and innovation. Open futures are those where more than one direction or course of action is feasibly possible; from the options available a choice has yet to be made. The notion of "futures past" refers to situations where previous opportunities to construct a given future were passed over. When such situations are recalled in the present, memories of the future are brought into play (Adam and Groves, 2007, pp. 196–200). When anticipated futures come to pass and are documented over time, relevant accounts are often infused with the idea of progress. However, if anticipated futures regularly fail to transpire then a temporal sense of cyclicity may prevail.

Temporal understandings are connected to the rhythms and cycles of biological nature, climate, and the cosmos, including the moon, sun and solar system. These connections are mediated by our senses, emotions and bodies. The scent of spring, for example, may recall a previous intimate association that triggers feelings of loss or hope. These are temporal as well as emotional dispositions. Calling upon our senses to note our breathing, the circulation of blood, beatings of the heart, movement of limbs, and the delivery of speech invokes the present-ness of lived temporality (Lefebvre, 2004, p. 21). Sensory, emotional, physical and intellectual interactions among people, underscored by natural rhythms and cycles, generate specific socio-cultural temporalities through craft, music, storytelling, mediated text, and mythic belief structures. Within the contemporary global context, socio-cultural temporalities are multiple and co-extensive. Thus, one's particular sense of temporal situatedness is imbued with a global dimension.

Worker experiences of employer-imposed systems of time management adversely affect the bio-rhythms of felt life within and outside the workplace. Unconstrained capitalist exploitation manifested in low wages, increased workflow and workplace surveillance, arbitrary alterations in shifts and working hours will stress the temporal sensibilities of workers and their families. This, in turn, will intrude upon the social solidarities of memory, in-the-moment rituals of present-ness and inter-generational future-orientedness which are constitutive of working-class culture.

Under global capitalism, the temporal repercussions of worker exploitation have increased in scale and severity. Worldwide deregulation of established labor laws under neoliberal policy regimes led to the implementation of irregular and flexible work schedules. A 2000 study of 21,505 workers employed in 15 European countries found that so-called "normal" or "standard" working hours were more the exception than the rule. Only 24 percent of the workers sampled routinely worked between 7:00–8:00am and 7:00–8:00pm, Monday to Friday (Costa et al., 2004, pp. 832–833). The detrimental temporal effects of night and shift work have been documented in studies pertaining to the correlation between work schedules, worker health and well-being, and to the internal synchronization of circadian rhythms (Fischer et al., 2004, p. 832; Birth, 2007). Meanwhile, the relocation of Western-owned assembly plants to Mexico, Southeast Asia and China has had profound micro-temporal consequences for local workforces and communities. Occupation/work-related diseases, workplace injuries and premature deaths were common occurrences. More recently, the imposition of flexibilized work practices and irregular shift schedules has become globally coordinated. As I will explain later, call center and ICT assembly work times in dispersed localities can be staggered across world time zones.

Coevalness

The co-extensity of epochal and temporal sensibilities brings us to "coevalness." This concept concerns intersubjective and reciprocal communication among individuals, groups, institutions and cultures with disparate apprehensions of time. This quality of communication is not pre-given; within prevailing structures of power, univocal assumptions about the nature of time may suppress other perspectives. In his historical critique of Western anthropological research, Johannes Fabian argues that anthropology constructed its object in ways that erased the historical and temporal perspectives of the cultures under investigation, a process he calls "the denial of coevalness" (Fabian, 1983, pp. 30–31). Such cultures are externalized as "other" to the assumed centrality of our time (the "other" is defined as backward, traditional, pre-modern, tribal, peasant, etc.). In opposition to this worldview, coeval research practices arrange or reorganize historical and epochal time in a way that gives identity to the "other" (in relation to oneself). Such practices advance the possibility of co-temporal understanding (with regard to otherwise disparate ways of life) and appreciate that temporal relations are embedded within global social totalities (Fabian, 1983, pp. 31–32). It is important here to realize that coevalness and conflict are not mutually exclusive. Recognizing the other's modus operandi may entail a reluctant acknowledgement of *their* worldview and *their* perspective on history. With the decolonization of Western empires after 1945 and the demise of Cold War politics after 1989, transnational, global-local, and trans-local cultural interactions created cosmopolitan social worlds. The Eurocentric preconditions for the denial of coevalness have been superseded by a vast mosaic of international, diasporic and ecumenical activity.

At the same time, however, the denial of coevalness has assumed a class dimension within global capitalism. Previously, Fordist-Keynesian political economies featured an inbuilt coeval tension between capitalists and organized workers. This situation is usually termed a class compromise. This may take the form of mutual cooperation; workers benefit from increasing wages and salaries and capitalists make profits from increasing aggregate demand. Alternatively, in a negative class compromise, two mutually opposed sides acknowledge that neither is strong enough to definitively prevail. The working class's associational power curtails the material interests of capitalists and vice versa. These forms of class compromise play out in the spheres of exchange (labor market regulation, demand management), politics (social democracy, corporatism), and production (union recognition, labor process co-management) (Wright, 2010, pp. 338–351). Class compromises can be regarded as coeval even if they are riven by tension and conflict. Each side recognizes the other's historical outlook, willingly or reluctantly. In these circumstances, bourgeois and working-class cultures co-exist. Global capitalism requires the erasure of this coeval relationship. Workforces are treated as a collateral or disposable component rather than as a fixture internal to capital-labor relations.

The worldwide configurations of this process have been outlined by Ann Hoogvelt. She observes that economic and social power relations constitute "a three-tier structure of circles," which "cut across national and regional boundaries" (Hoogvelt, 2006, p. 163). In this context:

> We may count in the core some 20% of the world population who are deemed "bankable" and therefore able to borrow funds. They are encircled by a fluid, larger social layer of between 20% and 30% of the world population (workers and their families) who labour in insecure forms of employment, thrown into cut-throat competition in the global market. State-of-art technology, frenzied capital mobility, and neo-liberal policies together ensure both a relentless elimination of jobs by machines and a driving down of wages and social conditions to the lowest global denominator. The third, and largest, circle comprises those who are already effectively excluded from the global system. This does not mean that they are not affected by the global system; on the contrary, they carry more than their fair share of its burdens, of environmental degradation and resource depletion, of war and conflict, of forcible dispossession. But what it means is that they are expendable.
>
> (Hoogvelt, 2006, pp. 163–164)

Those "already effectively excluded from the global system" live on the edges of paid work, precarious or otherwise. For the most part they form the burgeoning slums of African, Latin American and South Asian metropolises, such as Lagos, Cairo, Johannesburg, Mumbai, Karachi, Bangkok, Kolkata, and Sao Paulo. Myriad processes of micro-exploitation operate within this massive informal sector. Rag pickers, prostitutes, drug dealers, petty criminals, car washers, shoe shiners, street vendors lead lives of radical contingency. As Mike Davis has argued, they are not a reserve army of industrial labor to be employed and laid off as the economy grows and contracts. Rather, they constitute a surplus humanity produced by a dual system of structural exclusion (Davis, 2006). In the 1980s and 1990s, the standard International Monetary Fund (IMF), World Bank structural adjustment program of currency devaluation, privatization, tariff reduction, public sector contraction and food subsidy abolition benefited private lending institutions and transnational corporations at the expense of local working populations. Meanwhile, the profit imperatives of global agribusiness corporations undermined rural agricultural infrastructures. Throughout Africa and Latin America, fertile lands were converted into agricultural production units designed to service world commodity markets. For each agricultural commodity, the imperatives of corporate profit undermined rural livelihoods centered around ecologically sustainable mixed farming, local-regional market exchange and food security. The resulting dereliction of rural communities accelerated immigration to the cities (Davis, 2006; Li, 2011, pp. 281–298).

Time-related templates of worker exploitation

Within global capitalism, time reckoning, temporality, and the denial of coevalness are combinatory elements of worker exploitation. Together, they are constitutive of three integrally related templates: labor and labor time arbitrage; supply chain subcontracting and precarity; and the financialization of corporate performance. I will consider each in turn before sketching their combined result: a global time schema of capital-labor exploitation.

Labor and labor time arbitrage

The term "arbitrage" conventionally describes a particular type of behavior within global finance. Arbitrageurs make money by detecting minute and momentary price discrepancies across financial markets in order to buy cheap and sell dear. There is always the risk that the price of an asset will change between buying and selling it. When both trades can be transacted simultaneously, the result is risk-free arbitrage profit. Over longer time periods, transnational corporations profit from inter-geographic discrepancies in the cost of labor. In 2004, Morgan Stanley's chief economist, Stephen Roach, noted that a "new global labour arbitrage" was acting as a "powerful structural depressant" upon traditional sources of job creation in high-wage economies such as the United States, Europe and Japan. Three interconnected causal trends were identified. First, offshore outsourcing platforms, already established in manufacturing, were shaping the services sector from "low value-added transactions processing and call centres to activities with a high intellectual capital content, such as software programming, engineering, design, accounting, actuarial expertise, legal and medical advice and a broad array of business consulting functions." Second, Roach observed that ICT infrastructures had streamlined the logistics of supply chain management. In manufacturing the real-time monitoring of sales inventory, production and delivery trends became routine. For services, the intellectual capital associated with research analysis and consulting could be transmitted anywhere, instantaneously. Low-wage offshore knowledge workers could be seamlessly integrated into global service businesses, once the exclusive domain of knowledge workers in the developed world. Third, new imperatives of cost control brought global labor arbitrage to life. In the 1990s and early 2000s, wage rates in China and India ranged from 10–25 percent of those for comparable-quality workers in the United States and the rest of the developed world. Clearly, offshore outsourcing that extracted products and/or services from low-wage workers in the developing world was a routine profit-maximizing strategy. Meanwhile, in the United States and other developed economies, the outsourcing of jobs became the functional equivalent of imported productivity as global labor arbitrage substituted cheap foreign labor for more expensive domestic labor input (Roach, 2004, 2006).

Crucially, for our purposes, global labor arbitrage enables transnational corporations to make use of time arbitrage. The process is detailed in Shehzad Nadeem's analysis of India's outsourcing industries (Nadeem, 2009). On a global scale, transnational corporates exploit the time zone differences between different workplaces to construct a 24-hour business cycle. On a local labor process scale, time arbitrage entails the extension of work hours and the acceleration of work activities. Here, Nadeem details how, increasingly, severe systems of clock-time discipline are imposed upon workers. His fieldwork involved three companies that provided IT-enabled services or business process outsourcing services (in customer service, transcription, claims processing, graphic and content development). The other companies selected were an IT subsidiary of an investment bank and a major IT company (Nadeem, 2009, p. 22). Overall, two kinds of time experience were identified. Higher-end IT workers, usually salaried, worked long hours without overtime. Internal time schedules were elastic and breaks regular so long as a designated series of complex tasks were completed. In other words, time was stretched. These interviewees were representative of inter-networked global professionals working in complementary fashion across countries and time zones. Globally set deadlines for project completion would fall differently in different time zones. This would in turn differentially affect the diurnal cycles of individual workers in different locations (see also Sabelis, 2007). Call center workers involved in business process outsourcing experienced a quickened pace. Real time workplace surveillance allowed supervisors to set numerical targets, compare individual work rates, and restrict toilet and meal breaks. For these workers transnationally imposed clock-time discipline radically affects the temporalities and bio-rhythms of everyday life. Nadeem notes cases of night shifts disrupting family and social routines, and of work schedules curtailing attendance at religious festivals (Nadeem, 2009, pp. 32–33). Night shift workers experience the collapse of the work-life boundary. This situation does not resemble the socially embedded, pre-industrial work patterns described by E.P. Thompson. Rather, as Nadeem points out, "life is subsumed by work" (Nadeem, 2009, p. 35).

Supply chain subcontracting and precarity

Global labor arbitrage necessarily requires that work be subcontracted within and across transcontinental supply chains, but this is not a phenomenon that can be simply grid-mapped around the planet. Subcontracting arrangements are highly differentiated across economic sectors, geographic localities and socio-cultural lifeworlds. Let us now consider the overlapping patterns of exploitation involved from a time-related perspective. Most obviously, subcontracted work is insecure or precarious. The associated condition of precarity refers to insecurities of income, working hours, workplace health and safety, collective representation, legal rights, and skill advancement (Standing, 2011, p. 10). Precarious work has proliferated throughout developed, developing, and the poorest regions within construction, manufacturing, ICT assembly, consumer service and state sectors. Clearly, precarity is not just a manifestation of supply chain contracting; it also refers to a generalized condition of work. With regard to supply chain contracting as a time-related template of worker exploitation, one general feature of precarity deserves emphasis. Dependence on precarious work is a phenomenological condition. The absence of a securely based work identity has temporal ramifications. In precarious jobs, the low probability of personal progress or career advancement breeds short-termism and the incapacity to think beyond the immediate future. This affects interpersonal temporal experiences and outlooks. If one's position in life is unstable, social trust becomes fragile and contingent. Thus, precarious workers, broadly speaking, may have a weakened sense of intergenerational social memory. Shifting between jobs and localities undermines community belonging (Standing, 2011, pp. 18–25).

In given local-regional contexts, precarious work is embedded within an informal economy which operates in accordance with global supply chains. For example, Alessandra Mezzadri's investigation of the Delhi export-oriented garment sector found that traditional social structures central to the Indian informal economy had been restructured to supply international buyers from transnational fashion brands and retail chains (Mezzadri, 2008, pp. 603–617). The local nexus of this supply chain was observed to be pyramidal. At the top, merchant and manufacturing exporters have direct contact with the international buyer. Delhi-based processing centers cut, thread, stitch, iron, check and pack the garments. Subcontracting is used to increase tailoring capacity. Traditionally, such activities were performed by the *darzis* tailoring caste of northern India. From about 1980, increased export requirements loosened this arrangement as new workers arrived. Some 45,000 were then employed in the Delhi garment industry. By 2005, according to official sources, over

100,000 garment workers were employed, mostly in unregistered sectors. Mezzadri's research indicates that the actual workforce was two or three times larger (Mezzadri, 2008, p. 609). Ninety percent of garment factory workers were male migrants from the poor, outside states of Uttar Pradesh and Bahari. They obtained temporary and casual jobs, often on a daily basis. Inside Delhi's garment units, subcontracting allowed merchant and manufacturing exporters to maintain product and workflow flexibility in response to the needs of international buyers. Exporters also hire labor contractors or *thekedaars* to bring groups of workers to factories. Mezzadri notes that "in this way, the exporter is discharged of any obligation towards labour which is not 'his'" (Mezzadri, 2008, p. 609). The Delhi garment industry also makes use of ancillary non-factory labor, especially for the tasks of embroidery. These tasks are usually carried out by machines or by hand. Embroidery workers skilled with handlooms (*adda*-work) are supervised by proprietors in designated units. *Moti*-work or beading is realized through networks of home workers. Vendors and sub-vendors bring orders from Delhi to the proprietors of embroidery workers in outlying towns and villages. In this and other cases, family and socio-cultural temporalities are shaped by a local-regional economy that has become a node of transnational labor exploitation.

Supply chain subcontracting may also entail extra-economic coercion of precarious workforces and surrounding communities. Such is evident in the establishment of migrant labor regimes. Thus, Ching Kwan Lee and Yelizavetta Kofman note that "the state, sometimes in collaboration with international financialized institutions, deliberately creates a precarious workforce without full citizenship protection, in order to sustain a competitive export niche in the global economy" (Lee and Kofman, 2012, pp. 388–408). Such is exemplified in the demarcation of export processing zones by Southeast Asian governments (e.g. Indonesia, the Philippines, Singapore).

Perhaps the most vivid case, however, is Dubai, a Gulf state where prosperity and opulence depends on an "indentured invisible majority" (Davis, 2007, p. 64). In an essay entitled "Sand, Fear and Money in Dubai" (2007), Mike Davis outlines a social pyramid in which the al-Maktoum sheikdom and its cousins sit atop a layer of leisured local Arabic speakers (15 percent of the population), well-off expatriate managers and professionals (from North America, the United Kingdom, Europe, India, Lebanon, and Iran), and a populous bottom layer of "South Asian contract labourers legally bound to a single employer and subject to totalitarian social contracts" (Davis, 2007, p. 64). Dubai's luxurious personal lifestyles are attended to by "vast numbers of Filipino, Sri Lankan, and Indian maids." The construction industry is carried by "an army of poorly paid Pakistanis and Indians." They work "twelve-hour shifts six and a half days a week, in the asphalt melting desert heat" (Davis, 2007, pp. 64–65). Trade union strikes and protests are illegal and workers are immediately deportable. Migrants and their conditions of life receive little media coverage. Laborers are housed in desert camp settlements and excluded from the world of cosmopolitan consumption (Davis, 2007, p. 65).

Less visibly, African mineral workers endure slave-like conditions to produce the raw materials necessary for the production of cell phones, laptops, light bulbs, and cars. The minerals are smelted, refined and enriched in China, Thailand, Malaysia and Indonesia for electronics markets. In the Democratic Republic of the Congo, many mines are controlled either by armed forces or rebel armies. Communications researcher Christian Fuchs argues that "the informational productive forces of capitalism that create digital media are to a certain extent coupled to the slave mode of production in developing countries in order to reduce labour costs and maximize profits" (Fuchs, 2013). In Central Africa, "the slave mode of production" derives from local traditions of debt bondage (to outside lenders) and time bondage whereby workers belong to their employers for a given part of the week (Fuchs, 2013). In this form of subcontracted labor, pre-modern legacies of hierarchical brutality mesh with the profit imperatives of a global supply chain.

Financialization of corporate performance

Labor and labor time arbitrage, along with supply chain contracting and precarity, are closely connected to the financialization of Anglo-American corporations. The organizational aspects of this connection and the repercussions for capital-labor exploitation have not been properly examined. The significance of this connection, however, has been succinctly outlined by political economist William Milberg:

> The globalization of production has clear implications for pricing, profits, wages, and investment at the level of the firm and these have supported the process of financialization. Pressures for financialization and increased short-run shareholder returns have in turn spurred greater globalization of production as firms have divested the less competitive aspects of their production or relocated parts of the production process in order to lower costs.
>
> (Milberg, 2008, p. 445)

Corporate financialization and supply chain subcontracting did not only support rates of profit (and therefore rates of exploitation). In developed countries, some of the profits were used to support financial speculation, mergers and acquisitions rather than the revitalization of added-value production (Smith, 2012, p. 21). Consequently, workforces were either collateralized or abandoned. Furthermore, the inherent volatility of financialized capitalism transfers risks from investors to workers. Thus, the 2008 financial collapse immediately reduced bank lending and credit lines to large-, medium- and small-scale firms. Cuts in production, massive lay-offs and increased unemployment drove down consumer spending, which forced all kinds of businesses to close. Reduced credit flows and falling aggregate demand in large Western countries diminished export returns throughout East

Asia. This, in turn, reinforced the general tendency for global capitalism to disengage from collective labor.

From the preceding discussion a time-related schema of global capitalist exploitation can be formulated. The executives and senior managers of transnational corporations utilize real-time networks of communication and information retrieval to extract lags of clock time from supply chain management and subcontracting arrangements. Correspondingly, the cost and reliability of labor pools worldwide are evaluated and compared. On this basis, workers are flexibilized, re-designated or dispensed with. Thus, the real-time calculations of corporates increase the severity of clock-time disciplines on workforces. It therefore appears difficult for precarious, poorly paid "global workers" to deploy the categories of clock time against their employers. The real-time tourniquet of labor pool and labor process management radically disrupts the bio-social rhythms and temporalities of working human beings. On a macro-scale, global labor arbitrage and the precariousness of work profoundly influences temporal outlooks and dispositions among workers, families, and communities.

In this context, my understanding of exploitation builds upon the premise that labor is contained within capital and that it exists as an internal contradiction for capital. It is this internal relation that shapes capital and labor's struggle around different manifestations of time (time reckoning, temporality coevalness) and over the capacity to use time (as a collective resource). Capital and labor can, in fact, be locked in coeval conflict as was the case within Fordist-Keynesian political economies. Since then, global capital has sought to deny and eradicate this coeval conflict, and the internal relation upon which it was based. As a collective entity, workers are positioned as out of sight and out of time. In this regard, one must acknowledge that cultural denials of coevalness within formal Western imperialism have been superseded by the inter-cultural transparency of contemporary globalization. Global capitalism, however, has recast this denial as a strategy of class rule, through supply chain subcontracting, the manipulation of informal economies, the financialization of profit evaluation and the production of surplus humanity.

Labor resistance and the reclamation of time

The global schema of worker exploitation is not a settled accomplishment. In an economically interdependent world driven by instant transfers of information and knowledge, workers from disparate socio-cultural backgrounds can, potentially, position the whereness of their lives in relation to other lived places of work. To the extent that pan-worker solidarities emerge, the structures of global capitalist power can be collectively identified. The employment strategies of transnational corporations can thus be rendered visible by organized labor and associated activists through websites, DVD films and social media. Class-conscious workers can also exploit the fragile cost structures of just-in-time global supply chains. Throughout China's industrial zones, there have been reports of labor shortages, rising wages, and demands for union representation (*China Labor Watch*, 2012; *China Labour Bulletin*, 2012). Against this background, such workers had the capacity to exploit the vulnerabilities of global production networks and supply chains (Rainnie et al., 2011).

Collective actions against Foxconn, a Taiwanese manufacturing company and subcontractor for Apple, provide a case in point. In 2012, metal-processing and assembly workers at Foxconn Taiyuan in north China's Shanxi province erupted in protest. They were manufacturing iPhone casings and other components for final assembly at a larger Foxconn complex in adjacent Henan province. A shift in production requirements from iPhone 4S to iPhone 5, and the associated speed-up to meet Apple's delivery, gave workers the leverage to take collective action. A subsequent riot involving tens of thousands of workers spread throughout the factory complex. They used cell phones to send real-time images to local media outlets. Although senior government officials intervened, the tumult caused shipping delays for Apple just as consumer demand for the new iPhone 5 was peaking (Chan et al., 2013). This course of events discloses the efficacy of simultaneous and synchronized actions against transnational corporate supply chains (Rainnie et al., 2011).

More generally, Göran Therborn has identified four potentially critical forces against global capitalism. These are: empowered indigenous peoples, especially in Andean America and India; landless peasants and informal workers who comprise the urban slums of Africa, Asia and Latin America; manufacturing workers in China, Bangladesh, Indonesia, and elsewhere in the developing South; and radical middle-class protests against financial elites, authoritarian regimes, and austerity programs (as have unfolded in Spain, Greece, Chile, Turkey, Brazil, and Western metropoles) (Therborn, 2014). There is considerable potential for these concurrent political forces to be synchronized. In Spain, for example, during May 2011, the Indignados movement called for mass occupations of urban plazas against the imposition of austerity policies and the tax-funded bailouts of Spanish banks. In Madrid and Barcelona, major encampments forged transnational links with participants from workers' movements in Argentina, Chile, Italy and Greece. Anti-austerity populism in Southern Europe coincided with student protests in Santiago, London and Quebec, and inspired the Occupy Wall Street movement (Castañeda, 2012; Foran, 2014). The occupation of New York's Zucotti Park and subsequent declarations against financial elites, nationally and worldwide, was inspired by the popular overthrow of Tunisia's leader Ben Ali in December 2010 and mass demonstrations in Cairo's Tahrir Square during January and February of 2011 (Kerton, 2012). In 2010, the emergence of strikes across Chinese factories coincided with protests by Bangladeshi garment workers and a general strike of public sector workers and mineworkers in South Africa. In India, workers occupied a major Hyundai factory and instigated go-slows against the German car parts supplier Bosch (Silver, 2014). One must also acknowledge that trade unions, non-governmental organizations (NGOs) and informal worker organizations have developed transnational solidarities. In February 2011, a General Council meeting of the International Trade Union Confederation (ITUC) adopted a resolution calling for a range of actions to support those in precarious and informal work worldwide. Furthermore, ITUC affiliates and regional organizations were required to advance the rights of precarious workers. In June 2011, the steering committee of the International Domestic Worker Network (IDWN) successfully lobbied for the introduction of the International Labour Organization's (ILO) Domestic Workers Convention at the International Labour Conference (Bonner and Spooner, 2011). These modest initiatives reflect the necessity of advancing alliance formations incorporating the formally and informally employed amidst the occupational spread of precarity.

There is not the space here to detail the extent to which concurrent events, activities and movements were synchronized via media coverage and social media technologies. Such developments do, however, provide the backdrop for advancing a set of time principles that can be deployed against global capitalist strategies of worker exploitation. Here, the workers' traditional struggle to preserve, recuperate or advance their control over workplace time reckoning remains vital. Rates of pay, hours of work, speed of workflow and stability of employment over chronological time are indicators of class struggle on a transnational scale. From a union perspective, control over time reckoning at the workplace must be informed by the principle of temporal autonomy; the right to ascribe individual and social meanings to the passage of time. Families and local-regional communities, built around waged and informal work, should have the capacity to sustain socio-temporal customs without employer disruption. Meanwhile, if organized and informal workers alongside associated activists from disparate locations, can transnationally synchronize their protests and political interventions, coeval reciprocities within the opposition movement will accommodate the disparate experiences of gender, language, ethnicity and religion. On this basis, the "nowness" of simultaneous and synchronized political events will open up collective forms of temporal awareness in contradistinction to global capitalism and its time schemas of worker exploitation.

Note

1 The broader notion of contemporary globalization is encapsulated by the idea of global modernity. This can be distinguished from an older Eurocentric modernity which was ideologically inherent to Western imperialism, colonialism and the presumption of civilizational progress (Dirlik, 2007, pp. 94–97; Osborne, 2013, p. 75). Colonized cultures in Africa, the Middle East, South and East Asia, and Pacific Oceania were not seen as shapers of their own modern history (whereby Western colonial projects were encountered as an external force). With the spread of decolonization after 1945 and the demise of Cold War geopolitics after 1989, modernity was de-Westernized. Internally differentiated Confucian, Arabic, Islamic, African, Japanese, Western, Hindu and other cultures mingled across the global stage.

Different cultures and associated nation-states operated under the economic, political, cultural and cartographic legacies of colonialism. These legacies, and their contemporary repercussions, played out within an internally differentiated global modernity which was facilitated by the growing density of transport, information and communication networks. Intercity jet travel, satellite television, the Internet and mobile telephony generated a multiplicity of intercultural collaborations among politicians, writers, dramatists, artists, musicians, religious leaders and academic researchers. Transport and electronic networks also underpinned everyday socio-economic connections within and between diasporic communities across the globe. In this context, global modernity is convergent with what Barbara Adam calls "the global present." Human beings across the planet have gained "the capacity to partake in distant events as they [are] happening" (Adam, 2004, p. 120). This account of contemporary globalization *of itself* is compelling yet deeply misleading. It brackets out the associated formation of global capitalism and the reconfiguration of worker exploitation. These matters are the subject of this chapter.

References

Adam, B. (2004) *Time*. Cambridge: Polity.

Adam, B. and Groves, C. (2007) *Future Matters: Action, Knowledge, Ethics*. Brill: Leiden.

Birth, K. (2007) Time and the Biological Consequences of Globalisation. *Current Anthropology*, 48(2), pp. 215–236.

Bonner, C. and Spooner, D. (2011) Organizing Labour in the Informal Economy: Institutional Forms and Relationships. *Labour, Capital and Society*, 44(1), pp. 126–152.

Bryan, D. and Rafferty, M. (2006) *Capitalism with Derivatives*. London: Palgrave.

Castañeda, E. (2012) The Indignados of Spain: A Precedent to Occupy Wall Street. *Social Movement Studies*, 11(3–4), pp. 309–319.

Castells, M. (1996) *The Rise of Network Society*. Oxford: Blackwell.

Castells, M. (2000) Information Technology and Global Capitalism. In Hutton, W. and Gidders, A. (eds) *On the Edge: Living with Global Capitalism* (pp. 57–75). London: Jonathan Cape.

Castells, M. (2001) *The Internet Galaxy*. Oxford, Oxford University Press.

Chan, J., Pun, N. and Selden, M. (2013) The Politics of Global Production: Apple, Foxconn and China's New Working Class. *New Technology, Work and Employment*, 28(2), pp. 100–115.

China Labor Watch. (2012) *Tragedies of Globalization: The Truth Behind Electronic Sweatshops*, July. www.chinalaborwatch.org/upfile/2011_7_11/20110712.pdf.

China Labor Bulletin. (2012) *A Decade of Change: The Workers Movement in China 2000–2010*, March. www.clb.org.hk/sites/default/files/archive/en/File/research_reports/Decade%20of%20the%20Workers%20Movement%20final_0.pdf.

Costa, G., Akerstedt, T., Nachremer, F., Baltieri, F., Carvalhais, J., Folkard, S., Dresen, M.F., … Silverio, J. (2004) Flexible Working Hours, Health and Wellbeing in Europe: Some Considerations from a SALTSA Project. *Chronobiology International*, 21(6), pp. 831–844.

Davis, M. (2006) *Planet of Slums*. London: Verso.

Davis, M. (2007) Sand, Fear and Money in Dubai. In Davis, M. and Monk, B. (eds) *Evil Paradises: Dream Worlds of Neo-liberalism*. New York: New Press.

Dicken, P. (2004) *Global Shift*. London: Sage.

Dirlik, A. (2007) *Global Modernity*. Boulder, CO: Paradigm.

Dyer-Witheford, N. (2010) Digital Labour, Species becoming and the Global Worker. *Ephemera*, 10(3/4), pp. 484–503.

Dyer-Witheford, N. (2015) *Cyber-Proletariat: Global Labour in the Digital Vortex*. London: Pluto.

Fabian, J. (1983) *Time and the Other: How Anthropology makes its Object*. New York: Columbia University Press.

Fischer, M.F., Rotenberg, L. and de Castro Moveno, C.R. (2004) Equity and Working Time: A Challenge to Achieve. *Chronobiology International*, 21(6), pp. 813–829.

Foran, J. (2014) Beyond Insurgency to Radical Social Change: The New Situation. *Studies in Social Justice*, 9(1), pp. 5–25.

Fuchs, C. (2013) Theorising and Analyzing Digital Labour: From Global Value Chains to Modes of Production. *The Political Economy of Communication*, 1(2). www.polecom.org.

Goria, C. (2014) *Invisible Hands: Voices from the Global Economy*. San Francisco, CA: McSweeneys.

Harris, J. (2012) Outward Bound: Transnational Capitalism in China. In Murray, G. and Scott, J. (eds) *Financial Elites and Transnational Business* (pp. 220–241). Cheltenham: Edward Elgar.

Held, D., McCrew, A. and Perraton, J. (1999) *Global Transformations*. Cambridge: Polity Press.

Hoogvelt, A. (2006) Globalisation and Post-modern Imperialism. *Globalisation*, 3(2), pp. 159–174.

Hoogvelt, A. (2010) Globalisation Crisis and the Political Economy of the International Monetary Disorder. *Globalisations*, 7(1), pp. 51–66.

Hope, W. (2009) Conflicting Temporalities: State Nation Economy and Democracy Under Global Capitalism. *Time and Society*, 18(1), pp. 62–85.

Kerton, S. (2012) Tahrir, Here? The Influence of the Arab Uprisings on the Emergence of Occupy. *Social Movement Studies*, 11(3–4), pp. 302–308.

Lee, K.L. and Kofman, Y. (2012) The Politics of Precarity: Views Beyond the United States. *Work and Occupations*, 39(4), pp. 388–408.

Lefebvre, H. (2004) *Rhythmanalysis: Space, Time, and Everyday Life*. (Elden, S. and Moore, G., trans.). London: Continuum.

Levy, D. (2005) Offshoring in the New Political Economy. *Journal of Management Studies*, 42(3), pp. 685–693.

Li, M.T. (2011) Centring Labour in the Land Grab Debate. *The Journal of Peasant Studies*, 38(2), pp. 281–298.

Mezzadri, A. (2008) The Rise of Neo-liberal Globalization and the "New Old" Social Regulation of Labour: A Case of the Delhi Garment Sector. *Indian Journal of Labour Economics*, 51(4), pp. 603–619.

Milberg, W. (2008) Shifting Sources and Uses of Profits: Sustaining US Financialization with Global Value Chains. *Economy and Society*, 37(3), pp. 420–451.

Munn, N. (1992) The Cultural Anthropology of Time: A Critical Essay. *Annual Review of Anthropology*, 21, pp. 93–123.

Nadeem, S. (2009) The Uses and Abuses of Time: Globalisation and Time Arbitrage in India's Outsourcing Industries. *Global Networks*, 8(1), pp. 20–40.

Osborne, P. (2013) Global Modernity and the Contemporary: Two Categories of the Philosophy of Historical Time. In Lovenz, C. and Bevernage, B. (eds) *Breaking Up Time* (pp. 69–88). Göttingen: Vandenhoeck and Ruprecht.

Oxford Illustrated Dictionary (1979), p. 282.

Rainnie, A., Herod, A. and McGrath-Champ, S. (2011) Review and Positions: Global Production Networks and Labour. *Competition and Change*, 15(2), pp. 155–169.

Roach, S. (2004) How Global Labour Arbitrage Will Shape the World Economy. *Globe and Mail*. ecocritique.free.fr/roachglo.pdf.

Roach, S. (2006) Globalization's New Underclass. www.morganstanley.com/views/gef/a rchive/2006.

Robinson, W. (2004) *A Theory of Global Capitalism*. Baltimore, MD: Johns Hopkins.

Roth, K.H. (2010) Global Crisis: Global Proletarianization-Counter-Perspectives. In Fumagelli, A. (ed.), *Crisis in the Global Economy: Financial Markets, Social Struggles and New Political Scenarios* (pp. 197–237). Los Angeles, CA: Semiotext.

Sabelis, I. (2007) The Clock-Time Paradox: Time Regimes in the Network Society. In Hassan, R. and Purser, R. (ed.), *24/7: Time and Temporality in the Network Society* (pp. 255–278). Stanford, CA: Stanford University Press.

Silver, B. (2014) Theorizing the Working-Class in Twenty-First Century Global Capitalism. In Atzeni, M. (ed.), *Workers and Labour in a Globalized Capitalism* (pp. 46–69). New York: Palgrave Macmillan.

Singh, K. (2000) *Taming Global Financial Flows*. London: Zed.

Smith, J. (2012) Outsourcing, Financialisation and the Crisis. *International Journal of Management Concepts and Philosophy*, 6(1/2), pp. 19–44.

Standing, G. (2011) *The Precariat*. London: Bloomsbury.

Therborn, G. (2014) New Masses? *New Left Review*, 85.

Thompson, E.P. (1967) Time, Work Discipline and Industrial Capitalism. *Past and Present*, 38, pp. 56–97.

Tsing, A. (2009) Supply Chains and the Human Condition. *Rethinking Marxism*, 21(2), pp. 148–176.

United Nations Conference on Trade and Development (UNCTAD). (2008) *World Investment Report*. New York and Geneva: United Nations.

Winseck, D. and Pike, R. (2007) *Communication and Empire*. Durham, NC: Duke University Press.

Wright, E.O. (1997) *Class Counts: Comparative Studies in Class Analysis*. Cambridge: University of Cambridge Press.

Wright, E.O. (2010) *Envisioning Real Utopias*. London: Verso.

Zorn, D. and Dobbin, F. (2005) Managing Investors: How Financial Markets Reshaped the American Firm. In Cetina, K.K. and Preda, A. (eds) *The Sociology of Financial Markets* (pp. 269–289). Oxford: Oxford University Press.

4 Freedom 2044?

The temporality of financialization and scenario planning in *Looper*

Simon Orpana[1]

> The turbulence that once could be safely exported to the peripheries of the globe has now returned to haunt the American heartland.
>
> (Cooper, 2010, p. 169)

A Canadian TV advertising campaign for life insurance, popular in the 1980s and 1990s, assuaged the worries of would-be retirees with the soothing promise of "Freedom 55." In these commercials, a friendly voiceover prompts: "Imagine you had the power to visit yourself in the future." A somewhat harried, middle-aged professional man is then magically transported from "the rat race" to an idyllic locale, such as a yacht or beach-side promenade, where a smug, older version of himself explains how a London Life insurance policy gave him the freedom to retire early and still comfortably provide for his family. Significantly, these commercials emerged in a decade when the Western, Fordist ideals of life-long employment with a single company guaranteeing stable retirement income as a reward for loyal service were succumbing to the pressures of globalization. Among the processes that challenged the ideal of secure futurity were neoliberal economic models that targeted retirement funds and many other forms of socialized wealth as sites for generating private profits. Not only were financial managers increasingly investing pension funds in volatile but potentially more lucrative corporate securities (instead of stable but lower-yield government bonds), but with mounting capital flight, corporate mergers and downsizing, workers in North America and Europe were often pressured into making "give-backs" and relinquishing large portions of retirement and other benefits in order to save their jobs.[2]

The leveraging of future-oriented, socialized wealth in service of unpredictable, short-term gains has only increased in scope and scale since the 1970s, a fact that became spectacularly evident in the sub-prime mortgage crisis of 2008, the global repercussions of which are still being felt years later. The austerity measures that many governments have since attempted to impose, even as giant bailouts are delivered to the same financial institutions whose practices initiated the crisis, reveals financialization as a mode of governance that, originating in complicated, abstract instruments for commoditizing risk, has dramatic material effects at the level of everyday life and culture (Haiven, 2013, 2014; Jameson, 1997).

Among these effects is a disruption of our sense of time whereby our relation to the future has become obscure, remote, almost unimaginable and yet startlingly immediate at the same time. While this lived and felt contradiction escapes easy representation, cinema can supply us with narrative fantasies that give concrete form to collective anxieties. Though these fantasies often work ultimately to reinforce the flawed systems of which they are symptomatic, approached critically they might also supply us with new ways of relating to and challenging the ideological apparatuses that structure our world.

In the wake of the breakdown of major Western banking institutions in 2008, we might imagine an unsettling science-fiction "remake" of the Freedom 55 commercials, one tailored to the ensuing climate of austerity, precariousness and uncertain futures. In this narrative, instead of traveling to the future, an older man returns to the present to give advice to his younger self. Further inverting the original scenario, the younger man now seems comparably well off: he drives a nice sports car, has a lucrative job, and appears to be well-insulated from the squalor and poverty that surround him. His older self, in contrast, is a hunted fugitive on the run from a powerful syndicate of international (perhaps even "inter-temporal") gangsters. He has lost his beautiful wife, who saved him from a life of dissipation and addiction only to be shot and killed by the same gangsters, just when the older man thought he had realized his own "Freedom 55" dream. Desperate to regain his fortunes, the man has used a time machine to return 30 years into his past to the year 2044, where he tries to coach his younger self and alter history so as to restore his future life and happiness.

This, of course, is the plot of *Looper*, a film written and directed by Rian Johnson and released in 2012, four years into the current global financial crisis. Resonating with the sense of "no way out" that governmental responses to the crisis have reinforced for many Western subjects, *Looper* ingeniously deploys the circular, Oedipal structure of many time travel plots, where a protagonist's attempt to escape or change fate becomes the very vehicle by which fate is brought to pass.[3] *Looper* complicates this structure by offering a solution to the paradox – a breaking of the loop – that allows for what the film presents as a new historical "event" to emerge. However, while *Looper* arguably taps into viewers' desire for freedom from the sense of what I have called "free market claustrophobia" (Orpana, 2014), purporting to offer a way *past* the determinations of economics and austerity in late capitalist societies, it simultaneously reinforces and reifies ideological structures that bind us to that same order. It is this double movement of promised freedom and (sometimes not so) subtle re-inscription that makes the film an apt vehicle to illustrate and critique the paradoxical modes of temporality generated by globalized finance capitalism.

The film's title describes a variety of assassin who, in the year 2044, works for a crime syndicate operating 30 years in the future. The gangsters of 2074 make use of illicit time travel machines to send people they want dead back in time where they can be disposed of with impunity (biometric tagging making it difficult to do so in 2074). Joseph Gordon-Levitt plays Joe, a young Looper with designs to eventually leave the corruption of city life and move to France. Unlike his other, more present-oriented gangster friends, Joe is thus a Looper with a sense of futurity and an exit strategy.

The Loopers' victims are prepared by the gangsters of 2074 for easy disposal: the time machine delivers them in 2044 kneeling, their hands tied behind their back, their head covered by a sack, payment for the job in the form of heavy bars of silver affixed to their backs. I read this image of wealth magically appearing from the future, literally *on the backs* of disposable subjects, as giving narrative form to the temporality of finance capital itself, which trades upon our collective future in ways that garner short-term profits from the longer-term rendering of people and resources disposable. Like many real-world subjects affected by the 2008 crisis, Joe quickly loses his own sense of distance from precarity early in the film. Desiring to leave no loose ends, the gangsters of 2074 eventually "retire" the Loopers they employ, sending them back in time to be dispatched by their own past selves: a practice known as "closing the loop." When Joe fails to do this, the escape of his future self into 2044 creates a temporal rift – a literal collapsing of the future into the present that reverses his good fortune, and which, I will argue, symbolically captures the real-world effects of neoliberal financialization on our collective sense of time and futurity.

This collapsing of future time is linked by *Looper*'s narrative to a further set of anxieties about the struggles of the United States to maintain its position of global dominance in an increasingly turbulent geopolitical context. These anxieties are registered in the film by a mysterious figure known as The Rainmaker, who, operating out of China, takes control of the crime syndicates of 2074. Joe as an older man has also relocated to China where, after losing his amassed fortune due to drug addiction and high living, he is saved from his life of dissipation by a beautiful, unnamed Chinese woman who becomes his wife. As I will show, the tension that *Looper* establishes between the Joe of 2044 and his older self of 2074 provides a narrative structure that represents, critiques, and yet ultimately reinforces the convoluted temporality of financialization. In locating the source of Joe's troubles, not in the economic system of global capitalism, but in a particular constellation of relationships between the United States and the global South, *Looper*'s attempt to address the turbulence that globalization effects on our lives and worlds thus illustrates popular culture's uncanny ability to condense nebulous, imperfectly understood social, economic and historical realities into fantasy images which are then "magically" – which is to say, ideologically – managed and "resolved" via narrative constructs. By grappling with the convoluted temporalities of financialization and globalization at the level of narrative, *Looper* addresses Western subjects' hopes and anxieties regarding the balance of global power in an age of austerity, even as it circumscribes our vision of possible responses in a way that perpetuates the status quo.

Combining Marxist and post-Marxist frameworks of cultural analysis will enable us to question this circumscription, and to ask how our future world might truly become more egalitarian and humane. As what Fredric Jameson calls "socially symbolic acts" (Jameson, 2002, p. 1), popular film provides a privileged window into the cultural fantasies that give shape to our hopes and fears, casting troubling, real-world content in disguises that can then be symbolically managed in service of maintaining hegemony (Jameson, 1979). Building on Jameson's ideas, Christopher Breu emphasizes that cultural fantasy also provides a site where real-world issues might be "potentially worked through, rather than merely repeated and acted out" (Breu, 2005, p. 19). Although the circularity of *Looper*'s time travel plot provides a vehicle for critiquing the contradictory temporalities created by globalized finance capitalism, I argue that the solution the film ultimately offers, far from constituting an actual "event" in the sense of a new horizon of possibilities for thought and action, is actually *regressive* in nature, returning to a nostalgic American past where white patriarchy can still imagine itself the guarantor of a more prosperous world future, rescuing folk and soil from the racialized Others represented in the film by the Chinese gangster syndicate. *Looper*'s nostalgic vision of American renewal is thus buttressed by a racist, Orientalist logic, offering a particularly clear example of how popular culture performs work akin to a form of social control called "scenario planning" that Western governments have developed to maintain global relations of domination and exploitation in the neoliberal era.

Into the loop: exploitation and bare life under biocapitalism

In the opening scene of the film, the young Joe is stationed next to a cornfield where he awaits one of his targets to materialize, from the future, upon a tarp spread out to facilitate easy disposal of the body. As he checks an antique pocket watch, Joe recites conjugations of the French verb *avoir*, announcing the film's central anxiety regarding time and its consolidation in the possession of money. "Having" plays a double role here, signifying both the money that is about to materialize affixed to the back of Joe's victim and the idea of "having time," as symbolized by Joe's anxious glances at the pocket watch. Though the victim, who materializes seconds later, is literally "out of time" (again, in two ways), he (or possibly she) simultaneously and literally embodies money. After promptly dispatching his victim, Joe cuts into the deceased's canvas jacket to expose shining bars of silver affixed to his back. The graphic identification of anonymous human victims from the future with hard currency, an image that is repeated many times, provides a figure that resonates beyond strict narrative logic to speak to the suppressed social and political anxieties, fears and desires that the film summons in viewers without naming them as such. To start deciphering this backgrounded content, we might identify these victims as "already dead," the term used by Eric Cazdyn to describe those subjects whom globalized capitalism has left for dead, but who have yet actually to die (Cazdyn, 2012).

The already dead are a product of what Cazdyn calls "bioeconomics," or the contemporary tendency of governments to cede to the logics of austerity, drastically reducing protections and programs, while unapologetically admitting that we simply "cannot afford" to feed, house and clothe everyone (Cazdyn, 2012, pp. 152–155; 2010, pp. 59–60). Under bioeconomics, forms of social wealth are privatized, restructured and often dismantled, throwing increasing numbers of people on their own limited resources to manage the effects of globalization, downsizing and precariousness. The biopolitical imperative identified by Foucault, whereby the state has the power to "make live or let die," thus serves a bioeconomic logic in which the "letting die" of populations through the reduction of essential services simultaneously produces profits for the owners and managers of these "streamlined" operations: short-term accumulation is leveraged against the long-term destitution and even liquidation of whole populations (Foucault, 2003, p. 241). These social relations contribute to a form of temporality that Cazdyn, borrowing from palliative care models, calls "the new chronic," in which we struggle to find ways to manage the symptoms generated by the current system without addressing its inherent, possibly fatal contradictions (Cazdyn, 2012, pp. 1–18). In the new chronic, future possibilities for a more egalitarian, humane society are foreclosed in service of prolonging a painful, profitable "already dead" present indefinitely. Bioeconomics and the new chronic support an ideological formation I call the *zombie imaginary* by which, rather than addressing political and economic problems at the level of collective decision making, these issues are projected onto the bodies of (often racialized) people whose management and disposability then becomes the seeming solution to our problems (Orpana, 2014). These biopolitical formations have their material supports in finance capitalism, which commoditizes risk in a manner that rewards short-term gain over long-term planning for social well-being, the net effect of which is the polarization of wealth between a wildly affluent minority and a growing multitude of poor.

Social anxiety over generalized precarity is registered early in *Looper* by the squalor and poverty omnipresent in the unnamed American city in which Joe operates. This city is ruled by a gangster, Abe, and his men, called "Gat Men" because of the long-barrel handguns they carry. Here, in the polarization of society between a minority of affluent consumers (the Loopers and Gat Men) and a growing population of the dispossessed, disabled and unemployed, we encounter a familiar trope in sci-fi films made during the neoliberal era, from the overcrowded tenements of *Soylent Green* (1973), to the tent city of *They Live* (1988), to the damp, teeming streets of *Blade Runner* (1982). Each of these films also contains a biopolitical critique, using an exceptional figure (exemplified by the replicants in *Blade Runner*) to dramatically illustrate fears about the indeterminacy of contemporary life, where even the most gifted among us might be reduced to a mere cipher, disposable and anonymous. This critique is presented in *Looper* by the train of victims being sent from the future, offering a particularly vivid illustration of what Giorgio Agamben calls "bare life," or human subjects who can be killed with impunity, and whose exclusion from protection founds and supports the dominant order (Agamben, 1998). Similar to the treatment of the replicants manufactured by the Tyrell Corporation in *Blade Runner*, the precautions taken by *Looper*'s gangsters attempt to minimize any risk of the disposable victims escaping their fate. This tight circularity provides a fitting figure for the way in which contemporary capitalism leverages systemic resources to maximize its own profits while distributing the risks that it banks upon downwards to subjects whose social mobility is drastically compromised. Yet these already-dead subjects can and do manage to thwart their exploiters: having nothing left to lose, like Bruce Willis's character in *Looper*, they have a revolutionary potential to break free and challenge the system to which they are subjected (Cazdyn, 2012, p. 12).

We might further read *Looper*'s neatly packaged, monetized living corpses as figures of "human capital" who register the colonization of affect, communication, leisure and relationships under the mutation of economics Christian Marazzi calls "biocapitalism" (Marazzi, 2011, pp. 49–64). Like Cazdyn's idea of bioeconomics, the analysis of biocapitalism allows us to map determinative links between the financial realm of abstract risk analysis and the everyday life of people and the planet. It is finance's ability to quantify material processes, qualities and relationships, shearing them from their embodied, emplaced context and abstracting them into a realm where they might be manipulated by highly trained (and politically unaccountable) experts that makes financialization such a useful tool for contemporary capital to reap unprecedented profits from the globalized economy (Li Puma and Lee, 2004, pp. 119–140). Whereas the older regime of industrial capitalism extracted profits through the unpaid time it appropriated from workers employed making physical products (Marx, 1990, pp. 321–329), financialization has developed methods for profiting from relationships, events, ideas, affects, and probabilities: here the time appropriated by capitalists is no longer limited to hours spent working in a factory, but can be extended across the whole of waking life. This happens in ways ranging from the blurring of lines between work and leisure in accelerated and increasingly fluid "creative economies" (McRobbie, 2002), to the way the bodies of organisms have become profitable for their genetic make-up rather than for any particular qualities or work they might enact. This shift from extracting surpluses via the production of commodities to the commoditization of whole "forms of life" constitutes an intensification of the biopolitical logic of capitalism, a reality manifest in diverse realms, from battles over the copyrighting of DNA to the way subcultural lifestyles are packaged and sold back to the people who initially invented them (Hardt and Negri, 2005, pp. 109–115). Marazzi usefully links this biopolitical intensification to the logic of financialization, a moment in the life of capital that Marx described as ancillary to commodity production, but one which has, in the last 40 years, become a central engine of the global economy:

> beginning in the '80s, the expansion of finance was the other face of extending the process of extracting and appropriating value over the entire society. Financialization and the cyclic crises that characterize it should in fact be interpreted in the light of the biopolitics of labor, i.e., the post-Fordist productive strategies in which one's entire life is put to work, when knowledges and cognitive competences of the workforce (the general intellect that Marx spoke about in his *Grundrisse*) assume the role played by machines in the Fordist period, incarnated in living productive bodies of cooperation, in which language, affects, emotions and relational and communication capacities all contributed to the creation of value.
>
> (Marazzi, 2011, p. 113)

Returning to *Looper*, a capitalism that profits from interpersonal exchanges of affect, information and language might seem to contradict the film's image of hooded, cuffed and depersonalized victims who appear for an instant only to be liquidated, making the relationship between the Looper and his mark purely dehumanized and instrumental. Yet this very reduction of the Loopers' victims to bare life captures the hidden essence of biocapitalism, where the human universe of qualitative, interpersonal and affective "externalities" upon which contemporary capital increasingly depends for the generation of surpluses are reduced to the abstract, quantitative anonymity of money on the one hand and to subdued, interchangeable, and ultimately disposable, exhausted subjects on the other.

Key to both of these operations is the appropriation and management of time. The strange temporality of the Loopers' victims being doubly "out of time" constitutes a mediating point whereby the world of commoditized affect and lifestyle, symbolized in the film by La Belle Aurore nightclub that serves as the hub of Abe's gangster empire, is deeply implicated in the temporal exile suffered by the figures of precarious life who populate the film. This exploitative manipulation of time can be mapped and illustrated by some forensic accounting of Abe's operation. Capital in the form of silver flows into 2044 from the future, literally on the backs of the destitute, whom I read as illustrating the manner in which biopolitical finance capitalism robs us of time and futurity in a much more intensive and pervasive way than was possible under industrial capitalism. The silver is then converted by the Loopers into spendable cash (decorated by Asian heads of state, a detail that will soon become pertinent). This is done at a kiosk in La Belle Aurore, enabling Abe to accumulate transferrable capital in the form of silver and gold, whereas the Loopers dissipate their pay in the inconvertible form of cash. The Loopers are thus not merely providing an important service for the crime syndicates of the future (disposing of future bodies), but also help launder (future) funds through Abe's operation in 2044.

We can imagine Abe investing his amassed capital in ways that might consolidate the holdings of the future gangster syndicate, taking a significant cut for himself in the process. A second source of revenue is the prostitutes and dancers Abe rents out to his clients (most of the latter being the Gat Men and Loopers). After trading their precious metals for cash, the men in Abe's employ return a significant portion of their funds to their employer, spending it at the company canteen, as it were. A telling scene involving Joe's visit to his favorite escort, Suzie (Piper Perabo), reveals how the interpersonal forms of comfort offered at La Belle Aurore are always dispensed "on the clock" as a service paid for with funds that ultimately originate from the liquidation of future populations. This detail is telling of the logic of finance capitalism, where even the most personal of exchanges are implicated in the globalized mechanics of time and money.

Even while the Loopers return their earnings to Abe via the nightclub, they also risk becoming bare life themselves. While Joe's lucrative profession insulates him from joining the legions of homeless and destitute, it cannot shield him from the inevitable moment when he must "close his loop" and gun down his own, future self. Because a Looper's marks are packaged so as to be unidentifiable, any of Joe's victims could potentially be *him*, making his every hit a kind of Russian roulette. The only other option, that of letting one's loop escape, leads to the Looper becoming a hunted fugitive, which is exactly what happens in the time stream modified by the escaped Older Joe. *Looper*'s gangster operation thus serves as a figure for the interdependencies of financialization and biocapitalism, where a seemingly endless parade of disposable "talent" is buttressed by a growing, even more disposable underclass, with significant fluidity between the two categories. Only superficially different from their victims, the Loopers are similarly "out of time": their hedonistic lifestyles are paid for with funds accrued, ultimately, by liquidating their own future selves.

(Running) out of future time: the temporality of financialization

The cinematic figure of human capital arriving from the future registers the real-world temporality of biocapitalism itself which, as became spectacularly evident with the sub-prime mortgage debacle and attendant global financial crisis, trades upon potentially calamitous futures in order to realize gross profits in the present. In *Life as Surplus*, Melinda Cooper poignantly describes this temporality as a capacity for commoditizing and thus consuming the future itself. Productive here is the trope of waste that Joe highlights by describing his profession as "taking out the future's garbage." Reinforcing Joe's cynicism, Cooper describes the difference between the older, Fordist-industrial mode of production and the financialization regime as lying "merely in their temporalities: while industrial production depletes the earth's reserves of past organic life (carbon-based fuels), postindustrial bioproduction needs to depotentialize the future possibilities of life, even while it puts them to work" (Cooper, 2011, p. 25). Cooper argues that the promissory, future-oriented nature of contemporary financialization has allowed it temporarily to transcend the limits of industrial capitalism. By promising to make the mounting refuse of modernity profitable at an unspecified future date, industries like biotech and big pharma allow for capital expansion even in an age of mounting environmental crisis, widespread unemployment, and depleted resources. What is actually being gambled upon are the possibilities for future human life (and perhaps any life) on Earth: by failing to address the systemic flaws that prevent capitalism from ever truly realizing its promise of benefiting *everyone*, biocapitalism remains a utopian project that exploits and even liquidates the present and future lives it purports to serve. It is this vicious circle that *Looper* registers with its paradoxical temporality, in the image of a vast crime syndicate whose conduits into the future deliver a surplus of corpses and capital for disposal in the present. In the guise of cultural fantasy, Looper thus describes a dominant form of temporality upon which finance capitalism depends as a generator of surplus. Building upon Cooper's descriptions of biocapitalism, we might call this form of temporality "future-exploitative": approaching the limits of terrestrial expansion, capitalism must intensify its subsumption of the Earth and its life forms in a manner that extends to colonizing the future itself.

Retirement troubles and global turbulence

Joe's attempts to extricate himself from the syndicate take the form of a struggle over competing visions of his own future retirement. Here, the real-world contradictions inherent in contemporary, neoliberal approaches to retirement as described by Carolyn Hardin resonate in an uncanny way with Joe's predicament. Hardin critiques the neoliberal forms of temporality underpinning 401(k) retirement packages that became popular in the 1980s as a means of transferring the risk and responsibility of saving for the future from employers to American workers. Whereas retirement for (mostly male) workers under the Fordist model involved accounts managed by companies, who thus assumed the risk and responsibilities involved with investing these funds, the 401(k) model has workers paying into a fund, often supplemented by their employer, which it is then the individual worker's own responsibility to invest and manage. This model of retirement, argues Hardin, contains a contradiction in that it requires a rational, future orientation while the lived "time-sense" of workers under neoliberalism is predominately structured by the "future-in-present temporality of financialization" (Hardin, 2014, p. 102). Hardin shows how the tension between these two temporal orientations is mediated by the products and narratives of a financial planning industry that caters to workers managing their own retirement savings. Much like Cooper's example of a utopian biotech industry, this relationship creates a "substitute temporal horizon" that collapses the future into the now, a superimposition of temporalities that gives subjects an illusory sense that "the future is knowable, attainable, and enacted in the present" (Hardin, 2014, pp. 105–106). Retirement under neoliberalism is no longer something one patiently anticipates, like the destination at the end of a journey; it is something one lives in an immediate way, by continually navigating – and grappling with – tumultuous and uncertain securities markets.

Evoking a mutation of the Freedom 55 commercials, *Looper* renders this wrestling with the future in narrative form as a struggle between the Joe of 2044 and his future self. By stashing a significant portion of his earnings in the convertible form of silver, Joe acts like a responsible, neoliberal subject personally managing his income so as to provide for his future. However, when Joe's older self appears and escapes, we encounter a narrative vision of exactly the kind of "collapsing of the future into the present" that Hardin articulates as complicating these neoliberal temporalities.[4] Central here is the role played by risk and uncertainty, what Cooper calls "turbulence" in contemporary financialization (Cooper, 2010).

Edward Li Puma and Benjamin Lee identify the abandonment of the post-World War II Bretton Woods agreement in the early 1970s as the failure of the first international attempt to politically control the circulation of capital, inaugurating a new era of geopolitical uncertainty (Li Puma and Lee, 2004, pp. 71–73). During the following decades, an aggressive neoliberalism asserted the unregulated market as the best way to manage geopolitical turbulence and economic development, making yet another capitalistic projection that has failed to produce the promised, generalized benefits. This same era saw the rise of finance capitalism as a generator of unprecedented levels of surplus wealth for an equally alarming minority of subjects.[5] Under financialization, devices like derivatives, hedge funds and credit swaps, originally designed to mitigate risk, have become the instruments of a system that both reinforces and profits from increased global uncertainty and turbulence. Furthermore, a fetishistic relationship holds between the phenomenal world and the abstractions of derivatives whereby mechanisms that purport merely to quantify global risk actually shape and *steer* future events, presupposing and thus perpetuating the very circumstances and structures that they pretend to only describe (Li Puma and Lee, 2004, Ch. 5). In the grips of a Western geopolitical strategy based on disequilibrium (rather than the paranoid equilibrium of the Cold War period), turbulence thus becomes a category that spans economic, material and environmental registers (Cooper, 2010, pp. 168–170).

Images that speak to the contemporary production and monetization of turbulence saturate *Looper*, from small details like the undulations of cream dissolving in a cup of coffee, to the distortions of perception effected by the drug that Joe habitually drops into his eyes, to the uncanny environmental disruptions produced by the telekinetic boy, Cid. As we shall see, the US struggle to occupy a controlling position at the epicenter of this turbulence (rather than reaching an egalitarian compromise that would address and calm turbulence itself) is the central concern to which the convolutions of *Looper*'s time travel plot ultimately refer. In this reading, the disruption of temporality effected by Old Joe's escape into his own past illustrates a return of the repressed reality at the heart of US hegemony: whereas the global banking system conceals, via a fetishized commoditization of risk, the way in which it pre-emptively steers away from egalitarian solutions towards global imbalance, *Looper*'s time travel plot illustrates the "mayhem" that Westerners believe might ensue were a more egalitarian alternative to gain traction. Global turbulence is made incarnate in the threat posed by Old Joe to the younger Joe's retirement plans. Much like a worker managing a 401(k) account, the value of Joe's holdings fluctuates according to his ability to respond to the threat posed by uncannily incarnate future possibilities. Similar to the process of reification Li Puma and Lee identify as underpinning derivatives, the complex risks Joe faces are collapsed and objectified into a figure of his own older self – an entity whose wily ungovernability continually thwarts the younger Joe's attempts to secure his own future.

Scenario planning and the financialized lockdown of possible worlds

A further, illuminating window onto this conundrum opens when we consider the historical development of scenario planning as a governmental strategy contemporary with, and reinforcing, the financialization regime. Pioneered by executives from the Shell and ELF Gas companies in an attempt to respond to the energy crises of the 1970s, scenario planning leverages multiple possible visions of the future in order to try to anticipate and even profit from the uncertain, unpredictable outcomes of complex global systems of relations. Like financialization, scenario planning has become one of the dominant strategies employed by corporations and Western governments to navigate contemporary social, political, economic and environmental turbulence (Cooper, 2010, pp. 171–172). Cooper's description of the possible worlds logic of scenario planning speaks to the convolutions of *Looper*, where multiple time streams collapse upon and wrestle with each other:

> If x were to occur, what world would we be living in? If x had occurred (or had not), what world would we be living in? As these discontinuous ramifications unfold, the spectrum of alternative futures is expanded beyond the logical possibilities of simple prediction, affording us a glimpse not only into the possible futures of the actual world but also into the proliferating pasts and futures of counterfactual worlds. Unrestrained by the principle of non-contradiction, these counterfactual worlds are able to entertain contingencies that would be mutually exclusive in any one actual world.
>
> (Cooper, 2010, p. 173)

Like finance capitalism, scenario planning is more than merely a predictive or descriptive strategy: it utilizes a pre-emptive temporality that actually *steers toward or away from certain outcomes*, creating the preconditions for future events under the guise of merely trying to anticipate them. Importantly, this is done by strategically *narrowing* visions of what is possible: by presupposing certain, historically contingent facts (like the persistence of capitalism) as constant across all of the possible worlds under consideration, scenario planning can manage the *effects* of capitalism's contradictions and inequities even while it perpetuates the system itself. This reduction of political and moral issues to the seemingly "abstract" or fictional world of crisis scenario management is the post-modern form biocapitalism takes in the age of financialization.

Linking scenario planning to the logics of financialization, Cooper describes the "pre-emptive power of denationalized money" to secure "the imperial privilege of the US dollar" (Cooper, 2010, p. 181). Just as the figure of *Looper*'s gangster syndicate captures the contradictions and hidden terrors of finance capitalism, the film's clever use of the time travel plot provides a graphic example of scenario planning in action. It is here as well that Joe's personal struggle to secure his future opens into considerations of geopolitics, linking issues of an uncertain future faced by American workers to larger anxieties regarding possible contenders to America's position as global hegemon.[6]

Asiatic city versus US heartland: Orientalism in *Looper*

In the version of future reality that he struggles to preserve, Old Joe is saved from his destructive lifestyle by his Chinese wife (played by Qing Xu), a beautiful, seemingly powerful and affluent woman whom Joe meets during his "retirement" in Shanghai. Instead of traveling to France, Joe has gone to China, where he burns through his retirement fund in a few years of high living. After a stint as a Shanghai gangster, Joe loses his hair and meets his wife, who remains unnamed and silent throughout her appearances in the film's flashback/forward sequences. However, Old Joe's happiness is brought to an end by the mysterious and powerful gangster known as The Rainmaker, whose minions, in the course of attempting to send Joe back in time to be liquidated, also shoot his wife in the stomach, killing her painfully. When, after traveling from 2074 to 2044, Joe escapes being killed by his own past self, it is because he is motivated by the love for his deceased wife, a picture of whom he carries in his antique pocket watch. Old Joe cleverly reasons that in 2044, if he were to hunt down and kill the child who grows up to be The Rainmaker, he would instantaneously be restored to 2074 and the arms of his no-longer-deceased wife, having removed the efficient cause of both her demise and his own temporal exile. After narrowing the identity of the young Rainmaker to three American children, Old Joe kills in cold blood what turns out to be the wrong child. At the same time, the younger Joe, seeking refuge from the pursuing Gat Men and suffering from drug withdrawal, takes refuge on the farm of Sarah (Emily Blunt), a plucky single mother who also happens to be rearing the gifted telekinetic boy Cid (Pierce Gagnon), the lad who *will* grow up to be The Rainmaker in the timeline that Old Joe has fled.

If the first half of *Looper* chronicles Joe's descent into debauchery – a corruption that, as we shall see, is only *seemingly* corrected by the influence of his Chinese wife – the second half presents *Looper*'s vision of "genuine" redemption inspired by the younger Joe's identification with Sarah and her son. This binary structure is reinforced by a shift in generic registers. While the first hour of the film largely unfolds in the key of post-modern noir inaugurated by such works as *Blade Runner* and deploying a similar pastiche of Orientalism, techo-futurism, and urban squalor (minus the omnipresent rain), the second movement shifts to American gothic colored by spaghetti western and taking a detour through vigilante action films of the *Die Hard* variety. Thus, as the turbulence experienced by the two Joes and their pursuers intensifies, the temporality of the film itself retreats into filmic genres celebrating hegemonic masculinity and the pastoral American country homestead, where the flinty Sarah struggles to provide a decent upbringing for her prodigy of a son. Joe and Sarah, it turns out, have much in common. Like Joe, Sarah spent her formative years living high and wild in the wicked city. Unable to change her ways, Sarah gives the baby Cid to be raised by her sister in the country. When the sister is killed by one of Cid's fits of telekinetic rage, Sarah returns to the farm, cleans up her act, and sets about giving her boy a decent home life. After Joe stumbles upon the farm, he ends up bonding with Cid, and Joe realizes that to secure his own future happiness he will have to prevent his older self from killing the boy.

We are now in a position to articulate the central ideological mechanism of *Looper*, where the film's patriarchal and Orientalist visions are tethered to a set of stereotypical binaries that structure the two oppositional "scenarios" that Joe/Old Joe struggle to realize. On the one hand is the corrupt, cosmopolitan city, the values of which are embodied by the fallen prostitute Suzie and her gangster employer Abe. The former provides a foil to the virtuous Sarah and her lifestyle of resilient, country homesteading (while another of *Looper*'s rare female characters, the diner waitress Beatrix, prefigures the virtuous, "down home" allure that attracts Joe to Sarah). Abe, I would argue, finds a counterpart in Old Joe's Chinese wife, the powerful, seductive matriarch who saves Old Joe from his life of addiction and despair. The looped temporality of financialization is thus associated with an urban cosmopolitanism that is explicitly feminized and Orientalized in the sense described by Edward Said: the East – China specifically – is here depicted as a dangerous seductress, the site of a delusional madness that will make otherwise brave and good Americans want to kill innocent children. In this coding, Joe's Chinese wife, rather than *curing* him of drug addiction, is actually the *personification* of Joe's addictive delusion, something that his younger self will have to overcome to successfully resolve his retirement issues.

It is not difficult to detect, in this image of the East as seductress, anxieties over America's financial imbrications in the Chinese economy, which has been investing a great deal of its surplus in US Treasury bonds, making the Central Bank of China America's largest creditor. This is not actually so alarming as it might first sound, since running a deficit is one of the ways the United States has been maintaining its position as the guarantor currency for global trade in the post-Bretton Woods era. However, while the United States' privileged position has historically allowed it to deflect much of the pain of adjustments necessitated by its own excessive debt accumulation, China has more leverage than past countries that have occupied its creditor position (Germany and Japan) because it does not rely on US military protection (Otero-Iglesias, 2013, p. 5). This is partly why critiques of US fiscal policies launched by the Chinese Central Bank in the wake of the global financial crisis could not easily be ignored, and caused ripples for the US dollar in currency markets. China's call for fiscal reforms that would allow them to diversify their financial holdings across other currencies than that of the US dollar could shift the tides of geopolitics in favor of the rapidly industrializing country, producing a new, and possibly more even, distribution of power between Western and BRIC nations (Brazil, Russia, India and China; Ho Fung, 2013). This possibility has provoked some Western commentators to delegitimize China's critiques and encourage them to adopt policies that would lessen the foothold they have secured against US hegemony.[7]

In a similar way, by casting Old Joe's wife as a beautiful but ultimately dangerous seductress, *Looper* directly addresses American fears regarding China's increasing leverage in geopolitical affairs. Continuing our audit of Joe's rapidly depreciating retirement funds, we are shown how, in his life as a Shanghai gangster, his bars of silver, now translated entirely into Shanghai currency, alarmingly evaporate in the futurist-Orientalist equivalent of an opium haze. The climax of the film offers a fantasy of restoration when, back in 2044, a van appears containing what is presumably the gangster Abe's entire capital holdings which Old Joe has appropriated after storming La Belle Aurore. When the van is overturned by a blast of Cid's telekinetic turbulence, the bars of precious metal are scattered across the fields of sugar cane, graphically signaling a return of the "appropriated" funds to US soil where Sarah will be in a position to reclaim them in the closing scene of the film.

We can see, then, how the film's fractured temporality resolves itself into two competing scenarios whereby the older and younger Joes struggle to realize a future promise of security that the film consolidates into the opposing images of the Chinese wife and Sarah. In one telling scene, Old Joe, having taken refuge (in 2044) from his pursuers in a sewer, is struggling to keep the memory of his wife alive in his mind. We have learned earlier that, as the younger Joe responds to the attempts of Old Joe to change the time stream of 2044, Old Joe's memory shifts as well: remembered events from his own (future) history become clear or obscured according to whether events in the newly altered time stream are making his original future more or less likely to come to pass. As Willis's character peers at the picture of his wife set inside his antique pocket watch, he repeats "the first time I saw *her* face" as a mantra, attempting to hold on to this memory (of a now uncertain, future event) even as young Joe's relationship with Sarah threatens to erase the chain of causation that allows Old Joe and his Chinese wife to meet. To highlight the two possible histories/worlds in tension, the scene repeatedly cuts between a shot of the photo in the pocket watch and the living face of Sarah as she slaps the younger Joe out of the coma precipitated by his withdrawal from the drug to which he is addicted. As these two sets of possibilities – old Joe's "original" memory traces and those being created by young Joe – struggle to realize themselves in Old Joe's mind, the film cuts to the young Joe undergoing Sarah's rough ministrations, which literally wake him up from the Oriental(ist) spell that has transfixed his older self.

Conclusion: still dreaming the fuzzy logic of finance

Mark Bould has cited this same awakening scene in support of intertextual reading practices that, like the characters of *Looper* suspended between competing visions of the future, supply for viewers a "third space" in which productive complications and contradictions mobilize the politically indeterminate nature of popular culture (Bould, 2014, pp. 278–280). My own reading, however, shows how indeterminacy can be recuperated in service of ideological concerns precisely by way of this messy multiplicity and incongruity. Like scenario planning, popular culture works in service of dominant interests by expanding our horizon into a set of imaginary possibilities that, through their unstated assumptions and omissions, simultaneously limits how we think about present and future possibilities.[8] *Looper* accomplishes this by first eliciting viewer buy-in with its astute critique of the temporalities and hidden avenues of exploitation inherent in contemporary finance capitalism: a situation whose injustices we *feel*, even when the structural and economic avenues of its realization largely escape our understanding. *Looper* then offers a narrative resolution to this crisis, but does so in a way that reduces our vocabulary of response to alternatives that reproduce and reify, rather than challenge or change, the same political-economic complex it purports to critique. I agree with Bould that what *Looper* effects is not a "crypto-fascist overwhelming of the subject that shuts down cognitive and critical responses" (Bould, 2014, p. 280). More cleverly and insidiously, the film *excites* these critical, cognitive responses in such a way as to recapture them in a larger apparatus reinforcing American, patriarchal and Orientalist strategies of hegemony.

By setting up the resilient single mother Sarah as the suitable, alternative consort to the delegitimized, Chinese lover of Old Joe, *Looper* plays upon dangerous nationalistic and racialized tropes, uniting folk, soil and sovereignty into a feminized, nostalgic image of the American country homestead. We will recall that the single mother Sarah has left the evils of her urban past behind, a fact masterfully registered in mime when she provocatively smokes a non-existent cigarette on the front porch of her house. This artful, seductive gesture interpolates the viewer into the ideological logic of the film, which offers the fantasy of depression-era homesteading as antidote to the hardship and austerity faced by a growing number of Western subjects. Here we might use the film's fantasy logic against itself, *reading Sarah and her farm as the mirage, rather than the Chinese wife*. In this tack, Sarah's miming of the cigarette can be read as a wink to the audience as if to say: *you might think that by returning to the old Fordist values of denial, thrift and hard work you can have your cake and eat it too, maintaining capitalist relations even as the system chokes on its own excesses, but the nostalgic vision on offer is no more satisfying than the make-believe cigarette I'm smoking.* The white, patriarchal fantasy offered by *Looper*'s resolution is, in fact, comparable with Germany's National Socialist pre-World War II strategy of returning to the "real" wealth of people and land in the face of economic hardship, a strategy that led to the Nazi excesses against whole populations that they ruthlessly reduced to bare life. Indeed, *Looper*'s choice of a Jewish name, Abe, as belonging to the figurehead of a trans-temporal financial conspiracy has particularly odious resonances with the long history of anti-Semitic propaganda.

To help dispel *Looper*'s dangerous mirage it behooves us to look at other scenarios than those the film offers to address financialization's turbulence. Several authors can help us here. Both Miguel Otero-Iglesias and Giovanni Arrighi see in the contemporary constellation the possibility of a more equal balance of power between China and the United States, leading, if not to the "world-market society" prophesied by Adam Smith, then at least to a more equal collaboration between the two world powers for preservation of the current, troubled system (Arrighi, 2007, p. 7; Otero-Iglesias, 2013, p. 10). This would require abandoning the kinds of scapegoating reinforced both by *Looper* and prominent US economic analysts, which try to blame China for its deft navigation of the US-led debt export economy that has allowed America to live beyond its means for the past 40 years. Here, however, the image of resolution and thrift that *Looper* makes attractive in Sarah's steely determination needs be challenged as well, both for the move of casting child rearing as a fundamentally female vocation, and for the way in which this fantasy of autonomy-through-thrift leaves out the roving bands of hobos and vagrants that mar the country idyll to which the film repairs in its second hour.

In actualizing a new future that allows the telekinetic prodigy Cid to remain in the American heartland where he can be raised by his mother, *Looper*'s climax forestalls the alternative scenario whereby Cid and his turbulent powers migrate to China, where he precipitates massive "vagrant purges," extreme weather conditions, and other catastrophes. Paralleling scenario planning's presupposition of key traits – like the survival of global capitalism – as constant in all possible worlds, *Looper* refuses to seriously consider an alternative in which the global propensity to turbulence embodied in the child Cid is removed altogether. This, indeed, is the future that Old Joe struggles – and fails – to actualize. With Sarah teaching Cid to harness his powers for "good," we might translate the film's resolution into a call for "embedded liberalism" in which the uncertainties of globalized finance are "tamed" by some kind of "super Bretton Woods" agreement. Such a scenario is only likely to succeed, however, through cooperation between the world's powers, a situation that would be better represented in the film by Cid having both American *and* Chinese parents.

Extending this metaphor further, one could argue that disciplining the unruly global market requires more than just the world's "parents" collaborating – it is, after all, these same errant guardians who have helped create the current global crisis. What is needed is a new solidarity of global, bio-inclusive labor, a collaboration of workers in both the affective and industrial registers that can address the present and future of our collective life. Only pressure from such an assembly has the potential truly to "school" and reform the errant, turbulent market that currently governs (poorly) the fates of billions. Looking one last time to *Looper*, we might imagine the film's three sidelined women, the waitress Beatrix, the escort Suzie, and the Shanghai sophisticate who takes Old Joe as her consort, as particularly good architects for such an organization, representing as they do the precarious, affective laborers under biocapitalism and the subjugated vitality of the world's "developing" countries.

Notes

1 Many thanks to Emily Hill and Susie O'Brien for providing helpful commentary on previous versions of this chapter.
2 See Blackburn on the use of pensions in financialization (Blackburn, 2006, pp. 51–59). Among other questionable practices of hedge fund management, Blackburn cites Teresa Ghilarducci's argument that pension funds were deployed in service of the very downsizing that forced employees to relinquish agreed-upon benefits: "the stewards of labour's capital used pension funds in speculative investment activity, which closed plants and strangled communities" (quoted in Blackburn, 2006, p. 52).
3 Other cinematic treatments of this theme include *La Jetée* (1962), *The Terminator* (1984), and *Twelve Monkeys* (1995).
4 Eric Cazdyn provides his own examples of this temporality in his description of stranger-owned life insurance and "revolution insurance" as two forms of hedging to have emerged in response to the 2008 financial crisis (Cazdyn, 2012, pp. 159–163).
5 "From 1983 to 1998, daily trading in currency markets grew from $200 million to $1.5 trillion … by the time of the Asian market crash in 1997, the annual value of traded derivatives was more than ten times the value of global production" (Li Puma and Lee, 2002, pp. 203–204). By 2012, the global derivatives market reached a value of 20 times the world economy (Jay, 2012).
6 In a treatment that reinforces the critique provided here, Susie O'Brien provides a postcolonial critique of scenario planning as a strategy that attempts to maintain imperialist and capitalist imperatives amidst the increasing turbulence and distrust of authority in our era of globalization (O'Brien, 2016).
7 Ever since Zhou Xiaochuan, the Governor of the People's Bank of China, issued his critique of the US banking system in 2009, US commentators like Paul Krugman (2009, 2013) have been emphasizing what they take (or hope) to be the weakness of the Chinese position. Other commentators, however, have pointed out the novelty and possible relative strength of China's situation. Hung Ho Fung usefully emphasizes neoliberal deregulation as the precondition to global imbalances structuring the China-United States dilemma, pointing out China's potential for export and investment in developing BRICS countries (Ho Fung, 2013, pp. 159–160).
8 This is the basic thesis advanced by Jameson (1979), for whom the initial "utopian" moments of critique in popular culture are then recuperated in service of reification of the dominant social order; and by Stuart Hall (2007), who points out that hegemony works not by *forcing* certain narratives on us, but by limiting the options we have available for decoding them. My reading is also informed by Žižek's (2008) descriptions of contemporary ideology as working "permissively" through ambiguity and enjoyment.

References

Agamben, G. (1998) *Homo Sacer: Sovereign Power and Bare Life* (Daniel Heller-Roazen, trans.). Stanford, CA: Stanford University Press.

Arrighi, G. (2007) *Adam Smith in Beijing: Lineages of the Twenty-First Century*. New York: Verso.

Bergman, R. (Producer), and Johnson, R. (Director) (2012) *Looper* (Film). Beverly Hills, CA: Endgame Entertainment.

Blackburn, R. (2006) Finance and the Fourth Dimension. *New Left Review*, 39, pp. 39–70.

Bould, M. (2014) Of Eight Oscillations and Several Messages Carved in Flesh: Spectatorship, Cognition, and Affect in Dredd and Looper. *Science Fiction Studies*, 41(2), pp. 258–283.

Breu, C. (2005) *Hard-Boiled Masculinities*. Minneapolis: University of Minnesota Press.

Cazdyn, E. (2010) Bioeconomics, Culture, and Politics after Globalization. In Rethmann, R. et al. (eds) *Cultural Autonomy: Fictions and Connections*. Toronto: UBC Press.

Cazdyn, E. (2012) *The Already Dead: The New Time of Politics, Culture, and Illness*. London: Duke University Press.

Cooper, M. (2010) Turbulent Worlds: Financial Markets and Environmental Crisis. *Theory, Culture and Society*, 27(2–3), pp. 167–190.

Cooper, M. (2011) *Life as Surplus: Biotechnology and Capitalism in the Neoliberal Era*. Seattle, WA: University of Washington Press.

Deeley, M. (Producer) and Scott, R. (Director) (1982) *Blade Runner* (Film). Hollywood, CA: The Ladd Company.

Foucault, M. (2003) *Society Must Be Defended: Lectures at the College de France 1975–1976* (D. Macey, trans.). New York: Picador.

Haiven, M. (2013) Metaphoric Wealth, Finance, Financialization, and the End of Narrative. In Dufrense, T. and Sacchetti, C. (eds) *The Economy as Cultural System: Theory, Capitalism, Crisis*. New York: Bloomsbury.

Haiven, M. (2014) *Cultures of Financialization: Fictitious Capital in Popular Culture and Everyday Life*. New York: Palgrave Macmillan.

Hall, S. (2007) Encoding and Decoding. In During, S. (ed.), *The Cultural Studies Reader* (pp. 477–487). New York: Routledge.

Hardin, C. (2014) Neoliberal Temporality: Time-Sense and the Shift from Pensions to 401(k)s. *American Quarterly*, 66(1), pp. 95–118.

Hardt, M. and Negri, A. (2005) *Multitude: War and Democracy in the Age of Empire*. Toronto: Penguin Books.

Ho Fung, H. (2013) China's Rise Stalled? *New Left Review*, 81, pp. 154–160.

Jameson, F. (1979) Reification and Utopia in Mass Culture. *Social Text*, 1, pp. 130–148.

Jameson, F. (1997) Culture and Finance Capital. *Critical Inquiry*, 24(1), pp. 246–265.

Jameson, F. (2002) *The Political Unconscious*. Abingdon: Routledge Classics

Jay, P. (2012, September 26). Quadrillion Dollar Derivatives Market 20 Times Global GDP. *Truthout*. truth-out.org.

Krugman, P. (2009) The Chinese Dollar Trap. *The New York Times*, April 2. nytimes.com.

Krugman, P. (2013) The China-Debt Syndrome. *The New York Times*, October 18. newyorktimes.com.

Li Puma, E. and Lee, B. (2002) Cultures of Circulation: The Imaginations of Modernity. *Public Culture*, 14(1), pp. 191–213.

Li Puma, E. and Lee, B. (2004) *Financial Derivatives and the Globalization of Risk*. Durham, NC: Duke University Press.

Marazzi, C. (2011) *The Violence of Financial Capitalism* (K. Lebedeva and J.F. McGimsey, trans.). Los Angeles, CA: Semiotext(e).

Marx, K. (1990) *Capital: A Critique of Political Economy*, 1. (B. Fowkes, trans.). Toronto: Penguin Books.

McRobbie, A. (2002) Clubs to Companies: Notes on the Decline of Political Culture in Speeded Up Creative Worlds. *Cultural Studies*, 16(4), pp. 516–531.

O'Brien, S. (2016) We Thought the World was Makeable: Scenario Planning and Postcolonial Fiction. *Globalizations*, 13(3).

Orpana, S. (2014) The Law and its Illicit Desires: Transversing Free Market Claustrophobia and the Zombie Imaginary in Dredd 3-D. *The Review of Education, Pedagogy, and Cultural Studies*, 36(4), pp. 298–319.

Otero-Iglesias, M. (2013) China's Dollar Trap: Who is Entrapped and Who is Entrapping? And what Role for the Euro? *Asia Papers*, 7(4).

Said, E.W. (1979) *Orientalism*. London: Routledge and Kegan Paul.

Žižek, S. (2008) *The Sublime Object of Ideology*. New York: Verso.

5 Imagined futures and exceptional presents

A conceptual critique of "pre-emptive security"

Liam Stockdale

Introduction

In the autumn of 2011, the crime drama *Person of Interest* debuted on North American network television. The plot centers on a reclusive software billionaire and a former Central Intelligence Agency (CIA) agent who collaborate to wage a campaign of vigilante justice on the streets of New York City. What is unique about the premise is that their vigilantism is explicitly *pre-emptive* in nature. Using a sophisticated computer program that uses "pattern recognition" algorithms and "state-of-the-art surveillance technology" to "identify people about to be involved in violent crimes," the pair "works outside of the law to … stop crime before it happens" (CBS, n.d.). According to the show's backstory, the computer program was initially designed to collect and collate detailed surveillance data for the purpose of preventing another 9/11-style terrorist attack; however, it is discovered that its predictive capacities can also be applied to instances of more conventional violent crime. This leads its creator to recruit an ex-CIA partner to act on this information and prevent the impending violent incidents from taking place. Through this narrative, *Person of Interest* casts the idea of pre-emption as something of a panacea for security problems, as the viewer is induced to accept the legitimacy of the protagonists' anticipatory vigilantism – despite its clear transgression of the rule of law and its basis on unverifiable information obtained through highly intrusive measures – precisely because it is able to prevent crimes *before* they occur. In other words, the series offers an understanding of (in)security governance in which a traditional reactive strategy of responding to incidents that have already occurred is inadequate, while a future-oriented strategy of responding pre-emptively to incidents that have not yet occurred is the ideal.

It is thus fitting that *Person of Interest* premiered within days of the tenth anniversary of the 9/11 attacks, since its narrative quite accurately reflects a dominant way of thinking about global security in the post-9/11 era. Indeed, perhaps the most notable development in this area has been the proliferation of strategies premised upon what can be called a "logic of pre-emption" (Anderson, 2010, p. 790; Massumi, 2007, p. 9). Under such strategies, anticipatory interventions are undertaken in the present for the purpose of protecting against potentially catastrophic dangers located in the unknown – and ultimately unknow*able* – future (Anderson, 2010, pp. 227–235). This inflects the global politics of (in)security with an explicitly temporal component, since the fundamental imperative is to "police the future by anticipation" through the anticipatory exercise of political power in the present (Bigo, 2007, p. 31). In this sense, the very idea of "security" itself is (re)conceptualized through a temporal rather than spatial lens, as it is identified with successfully acting on the future in precisely this way.

Of course, the idea of "security" has always been infused with a fundamentally temporal element. As Foucault reminds us, "the specific space of security refers then to a series of possible events … to the temporal and the uncertain" (Foucault, 2007, p. 20). It is thus important not to overstate the novelty of the future-oriented, anticipatory turn in the post-9/11 global governance of (in)security. However, a burgeoning scholarly literature has identified and taken an analytical interest in this trend, as illustrated by the extensive work in critical security studies that documents and theorizes the proliferation of risk-based strategies and pre-emptive rationalities in the global "war on terror" (de Goede and Randalls, 2009, pp. 859–878). This suggests that the emergent "temporalization" of global security constitutes a distinct phenomenon, and that its constitutive practices diverge from other security rationalities in important ways that require critical attention.

It is in this context that this chapter can be situated, as it is concerned with further unpacking what is at stake in the ongoing shift from primarily spatialized to more temporalized understandings of global security, and the concomitant move from reactive to pre-emptive modes of governance in this context. This chapter thus seeks to develop a conceptual critique of such practices, which I collectively term "pre-emptive security" (Sullivan and Hayes, 2010; de Goede, 2008, pp. 161–185).[1] Given the widespread proliferation of anticipatory approaches to (inter)national security problems, understanding the latter is key to any account of the evolving security dynamics of contemporary globalization. Moreover, the overtly temporal character of such anticipatory action, combined with the global scope of the ongoing "war on terror," suggests that the politics of pre-emptive security represents a crucial point at which the intersection of time and globalization is made manifest in the contemporary context, particularly as it relates to changes in the conduct of warfare. It is in this capacity that the present chapter speaks most directly to the wider concerns of this volume.

The remainder of this chapter will proceed in three sections. The first develops a detailed account of how the logic of pre-emption is manifested in the context of post-9/11 global (in)security governance, offering a critical exploration of precisely what pre-emptive security can be understood to mean in the present context. The second section considers what is politically at stake with the proliferation of pre-emptive security strategies and tactics. Here it is argued that the logic of pre-emption prioritizes the role of imagination in political decision making, which implies a potentially radical shift in the way political power is exercised. In particular, since it significantly enhances the discretionary authority of state decision makers, I argue that there is an originary conceptual link between the logic of pre-emption and a politics of exceptionalism, in which the rule of law and associated limits on executive power are undermined. The third section builds on these arguments, contending that they raise serious questions about the democratic legitimacy of pre-emptive security strategies. It is then also argued that the very notion of pre-emptive security itself can be seen as conceptually incoherent, since its constitutive temporalities compromise its capacity to deliver "security" as conceived under its own normative framework. Instead, it can be seen to construct the present – which is where any experience of security must occur – as inherently *in*secure by merely replacing the type of subjective precarity that it is premised upon mitigating with another. I conclude with a brief reflection on the continued prominence of pre-emptive security strategies – particularly among liberal democratic polities – despite a decade and a half having passed since the 9/11 attacks that precipitated the (re-)emergence of pre-emption, and the significant conceptual and political problems highlighted in the chapter's analysis.

Pre-emption and (in)security governance in the post-9/11 era[2]

A growing body of work has meticulously documented the emergence of pre-emption as a prominent approach to governing global (in)security in the post-9/11 era (Anderson, 2010; Cooper, 2006, pp. 113–135). While addressing a wide range of practices – including, but not limited to the indefinite detention, extraordinary rendition and targeted killing of suspected terrorists; the widespread biometric monitoring of increasingly mobile populations; the pre-emptive detention of refugees and asylum seekers; the anticipatory freezing of monies and assets suspected of terrorist ties; and the so-called "Bush Doctrine" of pre-emptive interstate war – this literature has collectively highlighted the importance of temporally inflected, future-oriented strategies in the global security climate (Ericson, 2008, pp. 57–76; Butler, 2006; Mutimer, 2007, pp. 159–179; Kessler and Werner, 2008, pp. 289–308; Muller, 2010; Amoore, 2009, pp. 17–30; Isin and Rygiel, 2007, pp. 181–203; Weber, 2007, pp. 77–93; de Goede, 2012; Ehrenberg et al., 2010; Weber, 2007, pp. 109–128). The following discussion seeks both to tie together and build upon this literature's insights, so as to move toward a comprehensive conceptual account of pre-emptive security as it has (re-)emerged in the contemporary context. I hope also to begin to demonstrate how this emergent "primacy of pre-emption" entails significant, and potentially problematic, implications for the way political power is exercised with respect to questions of (inter)national security – a point which will be considered in greater detail in the next section (Massumi, 2007).

To begin, it is useful to consider how and why the logic of pre-emption has come to dominate the post-9/11 global security landscape. The key point here is that its rise has been largely coeval with the emergence of transnational terrorism as the dominant issue. This is because in the aftermath of the 9/11 attacks, which the prevailing narrative cast as an out-of-the blue irruption, the specific security problem of terrorism became identified with the broader existential problem of temporal contingency (Jarvis, 2009; Derrida, 2003, pp. 85–136; Adam, 2004, pp. 123–148).[3] From the outset, therefore, the post-9/11 governance of (in)security was effectively "temporalized," as the terroristic event was framed as ultimately "unpredictable in occurrence, characteristics, and effects" (Anderson, 2010, p. 228). Importantly, such a characterization rendered conventional global security logics such as deterrence or containment largely inapplicable with respect to transnational terrorism, since these are essentially ineffective in circumstances where the threat in question is defined by its radical incalculability (de Goede, 2008, p. 162; Sofaer, 2003, pp. 209–226). The global security environment thus became increasingly framed in terms reminiscent of Ulrich Beck's "risk society," with established practical certainties and techniques of control appearing to break down in the face of apparently novel threats (Beck, 2008, 1999; Anderson, 2010, p. 781; Kessler and Daase, 2008, pp. 211–232; Beck, 2002, pp. 39–55). Political authorities thus became "much less certain of whether and when they [were] secure, and how – and to what extent and at what price – security [could] be achieved" (Rasmussen, 2004, p. 382), and terms such as "extreme uncertainty" (de Goede and Randalls, 2009, p. 859), "radical contingency" (Dillon, 2008, p. 315), and other such expressions of perniciously excessive ambiguity came to pervade accounts of the global security environment. It is in precisely such a context that the logic of pre-emption proves compelling, since it is explicit about actively confronting uncertainty by enabling decision makers "to counter the unknowable before it is even realized" (Cooper, 2006, p. 120). Since the specter of transnational terrorism in particular appears to require precisely such a capability to be governed, the proliferation of pre-emptive strategies, mechanisms and technologies has been a central feature of the post-9/11 global security climate – to the extent that criminologist Richard Ericson has aptly referred to it as a "state of pre-emption" (Stern and Wiener, 2006, pp. 393–447; Ericson, 2008, p. 57).

The logic of pre-emption in this context is thus best understood as a political rationality that allows action to be taken in the face of apparent radical uncertainty. Normatively speaking, it is premised upon mitigating the potentially pernicious effects of that uncertainty by (re-)asserting control over the unfolding of the future through anticipatory interventions in the present. Yet because the future can never be known absolutely, a certain "objective uncertainty" is at the core of the logic of pre-emption (Massumi, 2007, pp. 7–8). As a result, the decisional rationality of pre-emption differs markedly from other logics of political decision making. Indeed, more conventional logics such as rational calculation and cost-benefit analysis – as well as other, more temporally inflected security logics such as deterrence – all assume at least some degree of certainty with respect to the substantive basis for action (Massumi, 2007, pp. 7–8; Sunstein, 2007). This is not to imply that these decisional logics require complete knowledge and/or total certainty, but rather that they are generally premised upon the use of verifiable knowledge to develop actionable assumptions with a certain degree of empirically rooted confidence. This diverges importantly from the logic of pre-emption, which takes radical *un*certainty precisely as its departure point, in that the mere potential of future danger provides the basis for anticipatory action in the present.

This location of radical uncertainty at the core of the logic of pre-emption gives corresponding decisions a highly speculative character. Indeed, the potential futures against which pre-emptive decisions are framed possess no ontological presence apart from their existence in the realm of the speculative. It is therefore not verifiable empirical fact that forms the primary basis of a pre-emptive decision, but conjectural articulations of potential (and potentially dangerous) futures (Aradau and van Munster, 2011, pp. 68–84). A pre-emptive security decision is thus effectively freed from any necessary link to verifiable knowledge or empirical fact (Aradau and van Munster, 2007, p. 108). Rather, decision makers are encouraged "to take into account doubtful hypotheses and simple suspicions ... to take the most far-fetched forecasts seriously," since the Cartesian "malicious demon" of the next terrorist attack might emerge at any time (Ewald, 2002, p. 288). Through its "appropriation of uncertainty as the *basis for* action" rather than an *impediment to* action, pre-emptive security creates a political climate in which decision makers are induced to "break free from the 'tyranny' of [the] plausible" (de Goede, 2008, p. 162; Kessler and Daase, 2008, p. 225). The indefinite future is precisely that which is being acted upon, and so any necessary tie to empirical fact would too heavily constrain the scope of decisional possibility, since no such "facts" exist in relation to a future that is beyond the realm of the wholly knowable. The logic of pre-emptive security thus resembles the so-called "precautionary principle" more familiar to theories of environmental governance, as the evidentiary circumscriptions of the decisional process are diminished considerably (Aradau and van Munster, 2007, p. 105; Ewald, 2002). Of course, pre-emptive decisions do not necessarily constitute entirely arbitrary exercises in speculation – although they can manifest as such. The point is rather that the role of the speculative and the conjectural *is* significantly higher than is the case under other political rationalities. Indeed, it must be, since the logic of pre-emption seeks to act upon a future that necessarily remains ultimately unknowable and thus "beyond the horizon of certainty" (Aradau and van Munster, 2008, p. 32; Amoore, 2009, p. 26).

This reliance on speculative and conjectural knowledge suggests that there are significant political consequences to adopting a pre-emptive approach to (in)security governance. In particular, it implies that a pre-emptive politics of security hinges on the exercise of the imagination by relevant decision makers, since this is the only way politically actionable "knowledge" about the ultimately unknowable future can be generated. The next section explores the implications of this point in greater detail, and specifically considers how such a prioritization of the imagination suggests a crucial conceptual link between the logic of pre-emptive security and the logic of political exceptionalism.

Imagination and exception: political implications of pre-emptive security

In considering the epistemological dimensions of the "politics of catastrophe" that animate the contemporary global security climate, Aradau and van Munster claim that the "imagination acquires epistemic primacy in relation to the unknown" (Aradau and van Munster, 2011, p. 85). Since pre-emptive security as discussed in this chapter is explicitly concerned with governing the inherently unknowable future, it follows that any pre-emptive decision will be premised to a significant degree upon the exercise of the imagination. Indeed, that "the logic of pre-emption prioritizes the power of imagination over the power of fact" is crucial to consider when thinking through the political implications of adopting a pre-emptive approach to (in)security governance (Salter, 2008, p. 243).

In this respect, perhaps the most important corollary is that, under a pre-emptive security rationality, anticipatory action can be taken against individuals or groups not merely on the basis of what they have done, nor even on the basis of what they have indicated through concrete actions or declarations that they will do, but rather on the basis of what the relevant authorities imagine that they might one day do (Weber, 2007, p. 115).

Consider the status of "unlawful enemy combatant" that was developed at the outset of the "war on terror" to be applied to suspected terrorists captured by American and allied forces. This specific designation was created in large part to enable the pre-emptive detention of individuals merely suspected of engaging, or planning to engage, in terrorist activities. As Ericson notes, a defining feature of this status is that detainees so designated may be held indefinitely on the basis of mere suspicion regarding "*imagined* future harm they *might* cause, rather than past crime" (Ericson, 2008, p. 63, emphasis added). In other words, the potential act for which they have been targeted exists only in a future constructed by the imaginations of the relevant decision makers. This archetypical example of pre-emptive security thus demonstrates that the imagination is not merely one element of the pre-emptive decisional rationality;[4] rather, it is *the* vital component, since it is imaginary constructions that fill the empirical knowledge void which invariably plagues any attempt to act on the ultimately unknowable future (Jarvis, 2009; Zehfuss, 2003, pp. 513–538; Edkins, 2002).[5]

The centrality of the imagination to pre-emptive decisions thus has significant implications for the way political power is exercised in the context of (in)security governance. In particular, it enhances the discretionary authority of state decision makers operating under a pre-emptive strategy. The key point here is that the "epistemic primacy" of the imagination ensures that any decision to intervene pre-emptively will be premised to a significant degree upon an evidentiary basis that is actively constructed by these figures. Again, this is because the future is ultimately unknowable, and thus can only be made actionable through the exercise of the imagination. Put differently, under a politics of pre-emption, the discretionary authority of those tasked with deciding when and how to act is radically augmented, extending beyond merely making the final decision to intervene to also include the prerogative to actively construct the epistemic basis for action by imagining a future against which anticipatory interventions can be mobilized in the present. In fact, this prerogative *must* be exercised if pre-emptive action is to be made possible at all, since the unknowability of the future upon which it is acting precludes the decision from being grounded in empirically verifiable fact.

Judith Butler captures these ideas well in a discussion of what she calls "deeming" in the context of post-9/11 (in)security governance (Butler, 2006, pp. 75–77; de Goede, 2012). Referring specifically to the indefinite detention of suspected terrorists – but describing a practice that is characteristic of pre-emptive security more generally – Butler emphasizes how a decision maker needs merely to "deem" an individual or group to be "dangerous" to trigger a potentially violent intervention such as indefinite detention (Butler, 2006, p. 76). As Butler puts it, "the decision to detain someone indefinitely is a unilateral judgment made by government officials who simply deem that a given individual or, indeed, group poses a danger to the state" (Butler, 2006, pp. 58–59). In other words, under a security regime whose primary aim is to pre-empt future acts of terrorist violence, "the 'deeming' of someone as dangerous is sufficient to make that person dangerous" (Butler, 2006, p. 59). The practice of deeming therefore constitutes "an unsubstantiated judgment that … works to pre-empt determinations for which evidence is required" (Butler, 2006, p. 76). This implies that the "dangerousness" of those who are "deemed" so exists only in the particular future actively imagined by the decision maker. An individual may thus be rendered subject to a violent anticipatory intervention such as indefinite detention simply "because some official 'perceives' a potential difficulty" (Butler, 2006, p. 76). In this sense, national security authorities are granted a "license to brand and categorize and detain on the basis of suspicion alone" (Butler, 2006, p. 76). The adoption of a pre-emptive rationality thus entails, in Marieke de Goede's words, "the transformation of evidentiary bases for security action," since it moves away from a grounding in verifiable fact and toward the realm of the speculative, the conjectural and the purely imaginary, and the task of creating this imaginary basis for action inevitably falls to those endowed with the authority to make security-related decisions (de Goede, 2011, p. 510). This enhances their discretionary subjectivity quite markedly, thus highlighting a key political implication of adopting a pre-emptive security strategy.

To better clarify these points through an illustrative example, consider the case of Mohamed Hersi – a Somali-Canadian man who was arrested by the Royal Canadian Mounted Police (RCMP) at Toronto's Pearson International Airport on March 29, 2011.[6] In what was described in the media as a "new kind of terrorist case," Hersi was detained prior to boarding a flight bound for Cairo via London and eventually became the first Canadian to be convicted of "attempting to participate in a terrorist activity" – an offence added to the Criminal Code of Canada in the aftermath of the 9/11 attacks (Freeze and MacArthur, 2011). Despite maintaining that he merely planned to study Arabic in Cairo for four months, the RCMP alleged that Hersi intended to go on to Somalia to train as a militant with the Al-Shabaab network – a group classified as a terrorist organization under Canadian law. Yet in a candid admission, the RCMP acknowledged that there was little concrete evidence that Hersi did, in fact, plan to train with Al-Shabaab as speculated (Freeze and MacArthur, 2011). This highlights the explicitly pre-emptive nature of the arrest, since the decision to act was nevertheless made in the face of the empirical knowledge deficit that characterizes anticipatory action. Moreover, the RCMP also admitted that its investigation "did not indicate that the suspect was a direct threat to his country or Canadians" at the moment of his arrest (Teotonio, 2011). In other words, the threat posed by Hersi that was pre-empted by his arrest did not exist at the moment he was detained. Indeed, not only was the evidence primarily conjectural, but Hersi himself was several degrees removed from actually posing an imminent terrorist threat to the Canadian state. The basis for his arrest therefore amounted to a series of speculative assertions relating to a potential future that may or may not have actually come to pass. That his arrest took place at all thus suggests that the RCMP made the decision to act based primarily upon the exercise of the imagination. Butler's formulation of "deeming" provides a good analogue for this case, as Hersi was "deemed dangerous" enough to be pre-emptively arrested, despite having committed no terrorist act at the time of his arrest, and with authorities openly admitting that he posed no immediate threat.

Mohamed Hersi was arrested on the basis of an *imagined future* threat that he *may have one day* posed had he ultimately succeeded in acting upon what he was *suspected* of *planning* to do. The case therefore illustrates both of the key points discussed so far in this section: that a pre-emptive security strategy requires decisions to be made through the exercise of the imagination, and that this vests within the relevant state authorities an enhanced degree of discretionary subjectivity. With respect to the first, the lack of verifiable evidence of guilt did not prevent the RCMP from acting, suggesting that it relied upon the imagination to construct a potential future against which Hersi's pre-emptive arrest could be framed in the present. With respect to the second, the fact that the RCMP could legitimately act upon such a speculative epistemic foundation shows how a mandate to pre-emptively govern terrorism significantly enhances the decisional discretion afforded to state security authorities.

This latter point is particularly important for understanding the broader political implications of adopting a pre-emptive security strategy. In particular, the enhanced discretionary authority granted to state decision makers by a reliance upon the imagination suggests an innate conceptual link between the logic of pre-emptive security and the logic of political exceptionalism. More specifically, it implies that pre-emptive security operates through a paradigm of political power resembling that which obtains in a so-called "state of exception."

As theorized by political theorists Carl Schmitt and Giorgio Agamben, a political "state of exception" obtains when the prevailing legal order is annulled and an arbitrary form of power, unconstrained by law and vested in a particular person or office emerges (Agamben, 2005; Schmitt, 2005, p. 12). "Exceptionalism" is thus characterized by two core components: the suspension of the juridical order, and a "decisionist" paradigm of political authority (Agamben, 2005, p. 23; Schmitt, 2005, p. 33). With respect to the first, a state of exception emancipates the highest form of legitimate political authority – usually the executive branch of a state government – from the circumscriptions imposed by the rule of law (Schmitt, 2005, p. 11; Lazar, 2006, p. 260). The executive's actions thus take a more arbitrary form, since the absence of any effective legal constraints invests it with an enhanced degree of decisional discretion. This logically implies the second component, which refers to a political condition in which the executive does not simply apply the law, but rather is effectively "the *source* of law" (Lazar, 2006, p. 257). Under a state of exception, therefore, every decision taken by a dominant authority is purer in its absolute singularity, such that, in Schmitt's words, it "emanates from nothingness" (Schmitt, 2005, p. 33). As is perhaps already apparent, such a paradigm of political power bears a conspicuous resemblance to that which the preceding discussion has associated with the logic of pre-emptive security. Indeed, the latter's prioritization of the imagination and concomitant expansion of discretionary authority suggests that it is operationalized through what amounts to a politics of exceptionalism.

To elaborate, consider first how the adoption of a pre-emptive security rationality presupposes the suspension of the juridical order. In this regard, the upshot of pre-emption's normative imperative to govern the unknowable future is that no imaginable potentiality can be *a priori* dismissed as a possibility that may have to be acted against.[7] This implies, more specifically, that no individual can be *a priori* absolved of suspicion in the present (Ericson, 2008, p. 66; Daase and Kessler, 2007, pp. 411–434). Indeed, a security regime under which the imagination constitutes the primary means of determining both the threat to be countered and the culpability of those to be targeted can become a regime of "universal suspicion that spells the end of innocence," where all individuals are placed upon a "continuum of risk" within the unlimited realm of the imaginary (Ericson, 2008, p. 66). Under such conditions, there is no clear way to prove an accused person's innocence once the imagination has been activated, since the accusation relates to an act that has, by definition, not yet occurred. In other words, an individual's demonstrable present innocence is rendered all but irrelevant when questions of security relate primarily to imagined futures that may or may not ever come to pass (Aradau and van Munster, 2007, p. 106).

Consequently, this undermines the very basis of the liberal democratic juridical order, since the collection of evidence and its evaluation via the appropriate judicial channels is rendered largely impossible when dealing with an imagined future event. Indeed, under the logic of pre-emption, the power to adjudicate an individual's (future) guilt or innocence – and thus also to subject her/him to potentially violent (present) interventions – will be transferred from the mechanisms of the juridical order to the whims of a designated political authority. The practice of pre-emptive security therefore brings about a condition where the established norms of a liberal juridical order not only *do not*, but in fact *cannot* apply, since they are simply incapable of accommodating the extended temporal horizons and attendant evidentiary uncertainties involved (Aradau and van Munster, 2009, p. 697). Prevailing legal norms are thus not merely suspended arbitrarily, but are essentially inapplicable when the logic of pre-emption informs security policy. The adoption of a pre-emptive approach to security questions thus suspends the juridical order almost *by default*, as questions regarding threat and culpability in the context of security become the purview of a designated authority.

Regarding the second component of exceptionalism, the particular type of decision required to operationalize pre-emptive security strategies suggests that the latter requires what amounts to a "decisionist" form of political authority to be made actionable. In this respect, recall the earlier point that the future potentialities against which pre-emptive decisions are framed possess no ontological presence apart from their existence in the realm of the imaginary. Again, this means that any decision to intervene pre-emptively must be premised to a significant degree upon elements that are actively constructed by the deciding authority. The purity of the decision required by the logic of pre-emption is thus quite marked, since it involves not simply the weighing of verifiable fact, but the active creation of conjectural knowledge about the future to serve as the primary basis for action in the present. Thus, precisely because the unknown future that is being acted upon constitutes something of an epistemic abyss, any decision to intervene pre-emptively "becomes in the true sense absolute," since only such a pure decision involving a degree of arbitrary, imaginative creativity can provide the foundation for action against a future that can never be known with certainty (Schmitt, 2005, p. 12). Just as it implies the suspension of the juridical order by default, therefore, pre-emptive security also enacts a decisionist form of political authority by what amounts to logical necessity. Indeed, its underlying imperative to govern the unknowable future can only be made actionable through the "pure" decision of an ultimate authority – a condition fundamentally associated with a state of exception.

As with the preceding points about the imagination, this link between pre-emptive security and political exceptionalism can be clarified through an illustrative example – in this case, the targeted killing of suspected Al-Qaeda operative and US citizen Anwar al-Awlaki, who was hit by a US drone strike on September 30, 2011, as his vehicle traveled down a rural road in northern

Yemen. This example is instructive, because the Obama Administration's decision to target al-Awlaki constituted both an explicitly pre-emptive exercise, and an archetypical example of political exceptionalism in practice. That the incident was an exercise in pre-emptive security is evidenced by the way that the broader CIA "Killing Program" under which it took place has been repeatedly framed in precisely such terms (Leander, 2011, pp. 2253–2268; Kessler and Werner, 2008). Moreover, the Administration sought to legitimate the killing after the fact by specifically invoking a pre-emptive imperative, asserting in a statement that al-Awlaki "posed some sort of imminent threat," the precise nature of which was not specified (Koring, 2011). This framing suggests that, much like the Hersi case, the decision to target al-Awlaki was based primarily upon an imagined future in which that ambiguous potential threat *had* actually manifested as the next attack. The incident also constituted a clearly "exceptional" act, since it exemplified both core components of political exceptionalism described above. In this regard, the killing was extrajudicial in that it was not authorized through conventional juridical channels or in accordance with prevailing standards of evidence. Rather, as critics have pointed out, it was ordered and carried out absent the due process of law guaranteed to all American citizens under the Fifth Amendment of the US Constitution. The juridical order was thus suspended at the moment the decision to kill al-Awlaki was made. Moreover, the *way* in which this decision was made was highly decisionist, since it accorded with the Obama Administration's broader approach to pre-emptive drone strikes that explicitly "concentrates power over the use of lethal US force outside war zones within one small team at the White House" (Dozier, 2012; Becker and Shane, 2012).[8] This executive-centered approach is archetypically decisionist in nature, as the final authority regarding who is to be killed, when the strike is to take place, and what counts as adequate evidence for a strike is effectively unconstrained by the rule of law and vested in the person of the President (Becker and Shane, 2012).[9]

The al-Awlaki case thus represents a useful illustration of the conceptual points outlined above, since it shows how pre-emptive security is operationalized through what amounts to a politics of exceptionalism. However, it also highlights the serious consequences of adopting a pre-emptive security rationality, particularly with respect to the relationship between state authorities and those being governed thereby. Indeed, the al-Awlaki killing shows how, even in liberal democratic states, a pre-emptive security strategy radically enhances the power of state authorities vis-à-vis citizens by minimizing the extent to which the law provides a normative mediation between them. In this regard, once the executive branch made the effectively arbitrary decision to pre-emptively target al-Awlaki, he could thenceforth be killed with impunity by agents acting on behalf of the executive, even though the letter of the law suggests that his citizenship ought to have insulated him from precisely such action. Moreover, al-Awlaki was deprived of the capacity to contest this precarious status, since the extrajudicial character of the targeting process insulates any such decisions from the judicial channels through which such contestation could be pursued. The US Constitution states that all citizens are protected against being "deprived of life, liberty, or property, without due process of law," meaning that an individual cannot be executed based upon a unilateral executive decision; yet al-Awlaki's killing – an archetypical manifestation of pre-emptive security in action – was authorized and carried out on precisely these terms. What the al-Awlaki case thus shows, is that the possibility of such action on the part of state authorities is a primary consequence of adopting a pre-emptive security rationality. In fact, the link between pre-emption and exceptionalism implies that maintaining such possibilities is a necessary condition for any viable pre-emptive security strategy.

These points begin to hint at the potential incoherence of such strategies, since it is questionable whether the resulting political condition is compatible with the idea of "security" in any meaningful sense. This point will be taken up on a more abstract level in the final section, but it is worth discussing here in the context of the preceding arguments. In this respect, the above discussion has shown how the logic of pre-emptive security demands that an effectively arbitrary, life-and-death decisional authority be granted to state security decision makers. This, in turn, creates an environment in which individual citizens are perpetually vulnerable to being "deemed" dangerous and thus subjected to a potentially violent anticipatory intervention. To be sure, pernicious excesses where wholly innocent individuals are targeted in this way are rare; however, the key point is that such instances *are* nevertheless *made possible* by the logic of pre-emptive security and its inherent link to the logic of exceptionalism. Indeed, beyond the Hersi and al-Awlaki cases, where the question of "innocence" is less clear, one need only consider two additional high-profile cases – the 2005 shooting death of Jean Charles de Menezes in a London Underground station, or the rendition and subsequent torture of Syrian-Canadian Maher Arar in 2002–03 – to appreciate this point (Vaughan-Williams, 2007, pp. 177–195; Taylor, 2006; Mutimer, 2007).

It is this immanent possibility that is problematic, however, since it signifies that even in states ostensibly committed to the rule of law and the upholding of human rights norms, the adoption of a pre-emptive security rationality always implies the potential for anyone to be arbitrarily deemed dangerous and thus subjected to state-sanctioned violence – whether at the airport, at the border, in a London Tube station, on a rural Yemeni road, or elsewhere (Amoore, 2008, p. 115). The implications for political subjectivity, and thus everyday life, in the democratic polities at the forefront of the pre-emptive governance of global security are thus quite significant – a point that is starkly evidenced by the pervasive embrace of such practices as pre-emptive detention, extraordinary rendition, and targeted killing by the security apparatuses of such putatively liberal regimes (Bigo and Tsoukala, 2008).

These potentially problematic aspects of pre-emptive security are exacerbated by the fact that its underlying imperative to actively confront radical uncertainty leads its praxis to favor the path of action over that of restraint. As former Vice President Dick Cheney asserted with respect to the potential threat posed by Saddam Hussein's Iraq, "the risks of inaction are far greater than the risk of action" (Daase and Kessler, 2007, p. 426). This proactive dimension ensures that adopting a pre-emptive strategy entails both a high level of activity, and a concomitantly increased likelihood that errors will be made. Indeed, as David Runciman asserts, anticipatory political strategies akin to pre-emptive security prioritize action over inaction and thus "do not take seriously enough the downside of getting things wrong." The result is that such incidents as the Menezes shooting can be easily framed as mere "accidents" or "mistakes" that, while regrettable, are an inevitable aspect of pre-emptive security (Runciman, 2004; Vaughan-Williams, 2007, p. 183). The blame can thus be placed on the unique circumstances of each case, obscuring the fact that – in a manner reminiscent of Virilio's notion of the "integral accident" or Derrida's idea of "autoimmunity" – it can actually be traced to the logic of pre-emption itself, since this is what ultimately makes such incidents possible (Virilio, 2007; Derrida, 2003).

In this sense, pre-emptive security's inexorable focus on the future offers a built-in justification for any violent excesses that may take place in the present. When so framed, it offers a helpful segue into the discussion of the inherent conceptual incoherence of pre-emptive security, and so provides a useful way to conclude this section. To elaborate upon this point, we must return to the question of temporality and recognize that the logic of pre-emption presupposes a very particular political relationship between present and future, in which the governance of the latter is prioritized while the former is construed in instrumental terms as the location of the anticipatory interventions required to make this priority actionable. This political temporality has a normative corollary, in that, taken to its logical conclusion, it implies that *any* action in the present can be cast as a legitimate means for achieving the overarching end of pre-empting a future catastrophe. In other words, it suggests that such "exceptional" acts as the shooting of Jean Charles de Menezes or the targeted killing of Anwar al-Awlaki can be legitimated for no other reason than that they take place in the present. Returning to the idea of exceptionalism more specifically, the logic of pre-emptive security thus inscribes the present as an exceptional temporal space where all anticipatory action – regardless of its extrajudiciality or arbitrary violence – is originarily legitimated. To put it another way, pre-emptive security is characterized by the enaction of a *temporalized state of exception*, in which the present is effectively "taken hostage" for the purpose of ensuring that the future unfolds in a particular way. Indeed, the exceptionalism of pre-emptive security is applied in specific relation to the dimension of time, with the present constituting the "exceptional space" where juridically unbound action takes place.

It is the widespread emergence of such a temporal exceptionalism that is at stake with the global proliferation of pre-emptive security rationalities in response to the threat of transnational terrorism. When considered alongside the illiberal paradigm of political authority that it also presupposes, this seriously calls into question the coherence of pre-emption as a security rationality, since any experience of security must take place in a lived present that is always already constructed as an inherently *in*secure "state of exception." This tension is particularly acute with respect to the self-identified liberal democratic states that are paradoxically at the vanguard of this pre-emptive turn, since the very principles that pre-emptive security strategies are mobilized to protect in the future are thus compromised in the present by the prosecution of a pre-emptive security strategy (Bigo and Tsoukala, 2008; Ehrenberg et al., 2010, pp. 512–565; Derrida, 2005, p. 99). These points suggests a more fundamental conceptual tension at the core of pre-emptive security that goes even beyond its problematic political implications and renders the very idea of pre-emptive security all but incoherent. The final section explores this point in more detail.

Toward a conceptual critique of pre-emptive security

In this last section, I seek to call into question the coherence of pre-emption as a security strategy from a purely conceptual perspective. The hope is that this analysis will supplement established lines of critique by illustrating how the problematic political implications discussed above are symptomatic of more originary tensions within the logic of pre-emption itself. The discussion will begin by continuing on the theme temporality, as the conceptual incoherence of pre-emptive security stems to a significant degree from its particular relationship to the question of time.

While the logic of pre-emption's appeal stems from its apparent capacity to "act on time" itself, it is important to recognize that it still presupposes a rather conventional understanding of time as a continuous flow in which past, present and future constitute discrete temporal spaces that can be affected by one another as time passes (Massumi, 2005, p. 5; Turetzky, 1998, pp. 152–153). In other words, the very idea of pre-emption is premised upon the Western/Enlightenment assumption of a linear temporality, as it relies upon corresponding cause and effect relationships through which action taken in the present will have a predictable effect on a future that will eventually become the present, and so on.[10] The conception of time that underpins the logic of pre-emption is thus one of continuous linearity, which assumes both a clear and enduring temporal ordering – in which the temporal spaces of past, present and future remain distinctly separate at any given instant – and a continuous temporal flow – in which the passage of time leads presents to become past, futures to become present, and new futures to emerge from the indefinite horizon of the to-come (Barrett, 1968, pp. 355–377; Mayo, 1968, pp. 275–291).

While fundamental to the logic of pre-emption, these temporal presuppositions destabilize its coherence as the basis for security governance mechanisms. To understand why, it must first be recalled that the pre-emption is premised upon a normative imperative that prioritizes the governance of the future above all else. Furthermore, as we saw in the previous section, the only way such an imperative can be rendered actionable is through anticipatory, potentially "exceptional" interventions in the present that produce a condition in which the present is effectively "taken hostage." The constitutive tensions within the logic of pre-emptive security begin to emerge when these points are considered alongside the fact that a linear temporality – which, again, is also presupposed by the logic of pre-emption – invariably leads what was the future to become the present, while also ensuring that a *new* future emerges on the temporal horizon (Kessler, 2011, p. 2178). This latter consideration is crucial, because this new future will inevitably embody the same unknowabilities and threatening contingencies as all futures, consequently requiring what is now the present to be in turn "taken hostage" for the purpose of governing these new future uncertainties. Thus, so long as the logic of pre-emption holds – and so long as time continues to flow – this process will continue *ad infinitum*. Indeed, the exigencies of linear time imply that an uncertain future will *always* be threateningly located on the temporal horizon of human subjectivity, while the normative imperative of pre-emptive security mandates that action be taken to govern this future through anticipatory interventions in the present. Thus, even if an impending catastrophe is successfully pre-empted, this will always be offset by the passage of time's inexorable production of a new, uncertain future whose own uncertainty will cast a shadow over the new present, requiring that it be subjected to another round of anticipatory action, and so on. The upshot is that the lived present – where human beings invariably exist and thus tangibly experience anything that might be understood as "security"[11] – can never be(come) fully insulated against the pernicious contingencies of the unknowable future (Turetzky, 1998, pp. 154, 159). This ultimately suggests that the praxis of pre-emptive security can never succeed on its own terms, since the actualization of its conception of security as protection from *potential* dangers is kept perpetually out of reach by the exigencies of the linear temporality it also presupposes.

To be sure, all security rationalities are ultimately incompletable processes in a similar sense, since "perfect security" – however it might be defined – is impossible. As such, this apparent flaw in the conceptual logic of pre-emptive security need not by itself compromise the latter's coherence as a security rationality. However, it embodies additional tensions that prove far more damaging in this regard. The key point in this respect is that pre-emptive security's capacity to deliver the form of "security" implied by its normative framework is seriously compromised by the components of its own constitutive logic. Put more simply, not only are the normative goals of pre-emptive security ultimately unrealizable, but it actually serves to bring into being the very type of *in*security that it is premised upon diminishing.

To explain how this is so, we must first consider how "security" is understood within the normative framework implicit in the logic of pre-emption. In this respect, recall from the introduction that adopting a pre-emptive rationality modifies the "conventional" meaning of "security" – as "a condition of being protected [and] free from danger" – by adding an explicitly temporal dimension (der Derian, 2009, p. 152). In particular, a pre-emptive approach aims to protect against *potentially forthcoming harms*, since the "threats" against which its associated action is framed are necessarily located in the future. Under the normative terms of pre-emption, therefore, "security" refers to the reduction – and, ultimately, elimination – of the precarious subjectivity caused by a vulnerability to *potential violence*. With this point in mind, a serious conceptual problem begins to emerge, which is that by responding to a threatening future by inscribing the present as an exceptional temporal space where state authorities can anticipatorily act on the basis of the imagination alone, pre-emptive security actually replaces the initial form of precarious subjectivity that it seeks to diminish – vulnerability to the threat posed by a radically uncertain future – with another form of precarious subjectivity – vulnerability to a state security apparatus that can act arbitrarily and violently in an attempt to govern those uncertainties. Put differently, the immanent potentiality of being targeted by such state action – of being "deemed dangerous" – is part and parcel of pre-emptive security, and this places individual citizens in arguably as precarious a position as does exposure to the potential catastrophes lurking in the future's unknowable depths. Pre-emptive security thus enacts precisely the sort of precarious subjectivity that it is normatively premised upon diminishing. Accordingly, its capacity to create a political condition congruent with its own understanding of "security" can be seriously questioned, and its coherence as a security rationality seriously compromised.

Now, it might be rejoined that even if such arbitrary violence is inherent to the logic of pre-emptive security at a purely conceptual level, this is of little importance since it has proven uncommon in practice. While empirically correct, this does not diminish the key conceptual point – which was also discussed in the preceding section – that such state action is necessarily *made possible*[12] by the logic of pre-emptive security and thus *can* occur anywhere at any time when the latter is in force. It is thus precisely this *potential* that is most conceptually problematic, since it illustrates how pre-emptive security replaces one form of precarious subjectivity with another. Indeed, it shows that in attempting to diminish the precarious subjectivity of exposure to the unknown future, the paradigm of political power required by the logic of pre-emption in fact brings into being an alternative precarious subjectivity in the form of immanent vulnerability to arbitrarily violent state action. This point is affirmed by Brian Massumi, who argues that a pre-emptive decision "strikes like lightning" because the imperative to act on the future requires that authorities be granted the prerogative to intervene as rapidly as possible whenever and wherever they see fit (Massumi, 2005, p. 5). The Menezes shooting and the al-Awlaki killing are again typical examples here. Moreover, as Judith Butler's discussion of "deeming" illustrates, this lightning strike of pre-emption can be both exceedingly violent and extraordinarily difficult to resist, since the futurity of the threat that the targeted subject is deemed to pose precludes any exculpatory demonstration of innocence in the present (Butler, 2006, pp. 75–77).

Accordingly, pre-emptive security inscribes precisely the opposite of that which it normatively promises, since the potential violence of vulnerability to the next terrorist attack is subjectively little different from the potential violence of vulnerability to indefinite detention at Guantánamo Bay, extraordinary rendition to face torture, death by drone strike in rural Yemen, or summary execution in a London Tube station. While such incidents may not be common, the point is that all individuals subject to a pre-emptive security regime are perpetually vulnerable to these sorts of actions. Indeed, a politics of pre-emption is ultimately premised upon precisely this possibility. It is this point that ultimately compromises the coherence of pre-emptive security. The latter is normatively premised upon protecting against the potential violence of an uncertain future, yet the mechanisms by which it does so create a lived present fraught with an alternative form of potential violence: the immanent possibility of an arbitrarily decided lightning strike by the state security apparatus. The application of pre-emption to the realm of global (in)security governance is thus incapable of delivering "security" as defined under its own normative framework. Instead, it brings into being a variation of the precarious conditions upon whose mitigation it is ultimately premised. Like so many of the governmental innovations of the post-9/11 era, therefore, the notion of pre-emptive security "ends up producing, reproducing, and regenerating the very thing it seeks to disarm" (Derrida, 2003, p. 99).

Conclusion

In a clear demonstration of what Agamben calls "security impos[ing] itself as the basic principle of state activity," the conceptual incoherence and manifestly illiberal character of pre-emptive security strategies has done little to hold back their continued global proliferation (Agamben, 2002). Although it is likely that the so-called Bush Doctrine of pre-emptive interstate war has been discredited by the imbroglio of Iraq and the subsequent global backlash, the idea of pre-emption remains a powerfully popular concept within the global security imagination, and continues to inform the prosecution of the broader "war on terror" in myriad ways. This shows that even more than a decade after the catastrophe that expedited pre-emption's (re-)introduction into global security discourse, and several years after the departure of its vanguard proponents in the Bush Administration, the notion of pre-emptive security as considered in this chapter remains highly influential. Accordingly, the sort of comprehensive critical account that I have sought to offer here remains key to any broader critical understanding of the global security climate in particular, and the intersection of time and globalization more broadly.

In developing this account, however, this chapter has attempted to highlight the potential dangers of unproblematically embracing this form of (in)security governance on a global scale. In particular, the discussion has emphasized how pre-emptive security's reliance upon the imagination leads it to radically alter the exercise of political power in a manner that greatly enhances the arbitrary decisional subjectivity of security authorities. This inevitably runs counter to the norms of liberal democratic juridical orders, and instantiates a particular iteration of the "state of exception" that is articulated in explicitly temporal terms. It was therefore ultimately claimed that the notion of pre-emptive security itself is ultimately incoherent, since not only does it represent something of a fool's errand whose ends are rendered unrealizable by the very temporalities it presupposes, but these inherent tensions, when combined with the political exceptionalism it demands, lead it to reproduce the very type of condition that it is normatively premised upon mitigating. It is thus hoped that, in addition to improving scholarly understandings of the logic of pre-emption in its capacity as a globally influential security rationality, the conceptual considerations developed here may have the more practical effect of giving pause to those growing numbers of policymakers who see the key to global security not merely in the exertion of control over space, but in the governance of time itself.

Notes

1 This term refers to security strategies and practices based upon the logic of pre-emption, consistent with prevailing terminological practice in the discipline.

2 The focus of this chapter is limited to the logic of pre-emption as it has (re-)emerged in the general context of post-9/11 (in)security governance. However, the idea of pre-emption has long been central to both philosophical/ethical discussions of the use of force in world politics, and more formal juridical treatments of the laws of war. On the former, see Walzer, 2000, pp. 74–85; on the latter, see Fletcher and Ohlin, 2008, pp. 155–176. Still, the current incarnation of pre-emption is qualitatively unique in certain ways, primarily due to the location of what has been called "radical uncertainty" at the core of its rationality for action. See: de Goede, 2008, pp. 164–65; Dershowitz, 2006.

3 I am referring here to the idea, posited by certain philosophers of temporality, that confronting the immanent uncertainty of being in time constitutes an originary struggle of the human existential condition.

4 Indeed, because they are made by human subjects, who are devoid of complete omniscience, the imagination necessarily plays *some* role in every political decision, pre-emptive or otherwise.

5 In discussing the importance of the imagined future in this context, it should be noted that particular renderings of the *past* also play a key role in influencing the sovereign imagination. Consider, for instance, how the invocation of an idealized past and the imagined return to a fictive state of "normality" is often used to legitimate an exceptional politics in the present. On these points, see Jarvis, 2009. Moreover, the role of trauma and memory cannot be discounted when considering the content of the imagined constructions of the future, as memoraic traces of traumatic experience can certainly influence the imaginary knowledges required for pre-emptive decisions. On this point, see Zehfuss, 2003, pp. 513–538; Edkins, 2002.

6 This case is still before the courts. The following discussion is based upon information made publicly available at the time of writing (November 2012).

7 This point was discussed in the first section of the chapter which described how a pre-emptive security approach takes radical *un*certainty as the basis for action.

8 The functioning of this approach is chronicled in detail by a *New York Times* investigation published in May 2012, and is worth quoting at length to illustrate the degree to which the Obama Administration's drone warfare campaign takes a decidedly decisionist form: "Every week or so, more than 100 members of the government's sprawling national security apparatus gather, by secure video teleconference, to pore over terrorist suspects' biographies and recommend to the president who should be the next to die. This secret 'nominations' process is an invention of the Obama administration, a grim debating society that vets the PowerPoint slides bearing the names, aliases and life stories of suspected members of Al Qaeda's branch in Yemen or its allies in Somalia's Shabab militia … [N]ames go off the list if a suspect no longer appears to pose an imminent threat … The nominations [then] go to the White House, where by his own insistence and guided by [chief counterterrorism advisor Jim] Brennan, Mr. Obama must approve any name" (Becker and Shane, 2012).

9 According to former Chief of Staff William Daley, Obama believes that his status as commander in chief requires that "his own judgments should be brought to bear on strikes" (Becker and Shane, 2012).

10 Indeed, pre-emption would be a blatantly incoherent idea absent these assumptions, since governing the future through anticipatory action in the present would be impossible if such causal temporal relationships did not hold.

11 The point here is that any experience of "security" must be primarily rooted in subjective perceptions, and, as philosophers of time have argued, "perception is always of something present … [and] … only grasps reality in the present moment." Accordingly, "security," however it might be conceived, can only be experienced in the present (Turetzky, 1998, pp. 154, 159).

12 It should be noted that the chief aim of this chapter is to critique the logic of pre-emptive security in the abstract. As such, the question of what is *made possible* under a politics of pre-emption is of greater importance than the empirical question of whether such potentialities are, in fact, playing out in contemporary security practices. It is important to bear this point in mind for the remainder of the chapter, as I am interested in expounding what is at stake with the rise of pre-emptive security in general rather than explicitly critiquing any specific practical manifestation thereof. Hidemi Suganami adopts a similar perspective in his critical interrogation of the concept of sovereignty, which provides a useful example of the type of argument from which I draw inspiration. See Suganami, 2007, pp. 511–530.

References

Adam, B. (2004) *Time*. Cambridge: Polity Press.

Agamben, G. (2002) Security and Terror. *Theory and Event*, 5(4), muse.jhu.edu/article/32641.

Agamben, G. (2005) *State of Exception*. Chicago, IL: University of Chicago Press.

Amoore, L. (2008) Consulting, Culture, and the Camp: On the Economies of Exception. In Amoore, L. and de Goede, M. (eds) *Risk and the War on Terror*. London: Routledge.

Amoore, L. (2009) Lines of Sight: On the Visualization of Unknown Futures. *Citizenship Studies*, 13(1).

Amoore, L. and de Goede, M. (eds) (2008) *Risk and the War on Terror*. New York: Routledge.

Anderson, B. (2010a) Pre-emption, Precaution, Preparedness: Anticipatory Action and Future Geographies. *Progress in Human Geography*, 34(6).

Anderson, B. (2010b) Security and the Future: Anticipating the Event of Terror. *Geoforum*, 41.

Aradau, C. et al. (2008) Security, Technologies of Risk, and the Political: Guest Editors Introduction. *Security Dialogue*, 39(2–3).

Aradau, C. and van Munster, R. (2007) Governing Terrorism Through Risk: Taking Precautions, (Un)Knowing the Future. *European Journal of International Relations*, 13(3).

Aradau, C. and van Munster, R. (2008) Taming the future: The Dispositif of Risk in the War on Terror. In Amoore, L. and de Goede, M. (eds) *Risk and the War on Terror*. London: Routledge.

Aradau, C. and van Munster, R. (2009) Exceptionalism and the "War on Terror": Criminology Meets International Relations. *British Journal of Criminology*, 49.

Aradau, C. and van Munster, R. (2011) *Politics of Catastrophe*. Routledge: New York.

Barrett, W. (1968) The Flow of Time. In Gale, R.M. (ed.), *The Philosophy of Time*. London: Macmillan.

Beck, U. (1999) *World Risk Society*. Cambridge: Polity Press.

Beck, U. (2002) The Terrorist Threat: World Risk Society Revisited. *Theory, Culture & Society*, 19(4).

Beck, U. (2008) *World at Risk*. Cambridge: Polity Press.

Becker, J. and Shane, S. (2012) A Measure of Change: Secret "Kill List" Proves a Test of Obama's Principles and Will. *The New York Times*, 29 May.

Bigo, D. (2007) Detention of Foreigners, States of Exception, and the Social Practices of Control of the Banopticon. In Rajaram, P.K. and Gundry-Warr, C. (eds), *Borderscapes: Hidden Geographies and Politics at Territory's Edge*. Minneapolis: University of Minnesota Press.

Bigo, D. and Tsoukala, A. (eds) (2008) *Terror, Insecurity, and Liberty: The Illiberal Practices of Liberal Regimes after 9/11*. London: Routledge.

Butler, J. (2006) *Precarious Life: The Powers of Mourning and Violence*. London: Verso.

CBS (n.d.). About: Person of Interest. www.cbs.com/shows/person_of_interest/about/ (accessed October 12, 2012).

Cooper, M. (2006) Pre-empting Emergence: The Biological Turn in the War on Terror. *Theory, Culture & Society*, 23(4).

Daase, C. and Kessler, O. (2007) Knowns and Unknowns in the War on Terror: Uncertainty and the Political Construction of Danger. *Security Dialogue*, 38(4).

de Goede, M. (2008) The Politics of Pre-emption and the War on Terror in Europe. *European Journal of International Relations*, 14(1).

de Goede, M. (2011) Blacklisting and the Ban: Contesting Targeted Sanctions in Europe. *Security Dialogue*, 42(6).

de Goede, M. (2012) *Speculative Security: The Politics of Pursuing Terrorist Monies*. Minneapolis: University of Minnesota Press.

de Goede, M. and Randalls, S. (2009) Precaution, Pre-emption: Arts and Technologies of the Actionable Future. *Environment and Planning D: Society and Space*, 27.

Der Derian, J. (2009) The Value of Security: Hobbes, Marx, Nietzsche, Baudrillard. In *Critical Practices in International Theory: Selected Essays*. London: Routledge.

Derrida, J. (2003) Autoimmunity: Real and Symbolic Suicides. In Borradori, G. (ed.), *Philosophy in a Time of Terror: Dialogues with Jürgen Habermas and Jacques Derrida*. Chicago, IL: University of Chicago Press.

Derrida, J. (2005) *Rogues*. Stanford, CA: Stanford University Press.

Dershowitz, A. (2006) *Pre-emption: A Knife that Cuts Both Ways*. New York: Norton.

Dillon, M. (2008) Underwriting Security. *Security Dialogue*, 39(2–3).

Dozier, K. (2012) Who Will Drones Target? Who in the US Will Decide? *Salon*, 12 May.

EdkinsJ. (2002) Forget Trauma? *International Relations*, 16(2).

Ehrenberg, J. et al. (eds) (2010) *The Iraq Papers*. Oxford: Oxford University Press.

Elmer, G. and Opel, A. (2006) Surviving the Inevitable Future: Pre-emption in an Age of Faulty Intelligence. *Cultural Studies*, 20(4–5).

Ericson, R.V. (2008) The State of Pre-emption. In Amoore, L. and de Goede, M. (eds) *Risk and the War on Terror*. London: Routledge.

Ewald, F. (2002) The Return of Descartes' Malicious Demon: An Outline of a Philosophy of Precation. In Baker, T. and Simon, J. (eds) *Embracing Risk: The Changing Culture of Insurance and Responsibility*. Chicago, IL: University of Chicago Press.

Fletcher, G. and Ohlin, J. (2008) *Defending Humanity: When Force is Justified and Why*. New York: Oxford University Press.

Foucault, M. (2007) *Security, Territory, Population*. New York: Picador.

Freeze, C. and MacArthur, G. (2011) Man Jailed in New Kind of Terrorist Case for Canada. *The Globe and Mail*, March 30,.

Isin, E. and Rygiel, K. (2007) Abject Spaces: Frontiers, Zones, Camps. In Dauphinée, E. and Masters, C. (eds) *The Logics of Biopower and the War on Terror: Living, Dying Surviving*. Basingstoke: Palgrave.

Jarvis, L. (2009) *Times of Terror: Discourse, Temporality, and the War on Terror*. London: Palgrave-Macmillan.

Kessler, O. (2011) The Same as it Never Was? Uncertainty and the Changing Contours of International Law. *Review of International Studies*, 37.

Kessler, O. and Daase, C. (2008) From Insecurity to Uncertainty: Risk and the Paradox of Security Politics. *Alternatives*, 33.

Kessler, O. and Werner, W. (2008) Extrajudicial Killing as Risk Management. *Security Dialogue*, 39(2–3).

Koring, P. (2011) Memo Equips Obama with "Licence to Kill." *The Globe and Mail*, October 9.

Lazar, N.C. (2006) Must Exceptions Prove the Rule? An Angle on Emergency Government in the History of Political Thought. *Politics & Society*, 34(2).

Leander, A. (2011) Risk and the Fabrication of Apolitical, Unaccountable Military Markets: The Case of the CIA Killing Program. *Review of International Studies*, 37.

Massumi, B. (2005) The Future Birth of the Affective Fact. *Conference Proceedings: Genealogies of Biopolitics*, 5.

Massumi, B. (2007) Potential Politics and the Primacy of Pre-emption. *Theory & Event*, 10(2).

Mayo, Bernard. (1968) The Open Future. In Gale, R.M. (ed.), *The Philosophy of Time*. London: Macmillan.

Muller, B. (2010) *Security, Risk, and the Biometric State: Governing Borders and Bodies*. London: Routledge.

Mutimer, D. (2007) Sovereign Contradictions: Maher Arar and the Indefinite Future. In Dauphinée, E. and Masters, C. (eds) *The Logics of Biopower and the War on Terror: Living, Dying Surviving*. Basingstoke: Palgrave.

Rasmussen, M.V. (2004) It Sounds Like a Riddle: Security Studies, the War on Terror, and Risk. *Millennium*, 33(2).

Runciman, D. (2004) The Precautionary Principle. *London Review of Books*, 26(7).

Salter, M. (2008) Risk and Imagination in the War on Terror. In Amoore, L. and de Goede, M. (eds) *Risk and the War on Terror*. London: Routledge.

Schmitt, C. (2005) *Political Theology: Four Chapters on the Concept of Sovereignty*. Chicago, IL: University of Chicago Press.

Sofaer, A.D. (2003) On the Necessity of Pre-emption. *European Journal of International Law*, 14(2).

Stern, J. and Wiener, J. (2006) Precaution Against Terrorism. *Journal of Risk Research*, 9(4).

Suganami, H. (2007) Understanding Sovereignty Through Kelsen/Schmitt. *Review of International Studies*, 33.

Sullivan, G. and Hayes, B. (2010) *Blacklisted: Targeted Sanctions, Pre-emptive Security, and Human Rights*. Berlin: European Centre for Constitutional and Human Rights.

Sunstein, C. (2007) *Worst Case Scenarios*. Cambridge, MA: Harvard University Press.

Taylor, P. (2006) The Terrorist Who Wasn't. *The Guardian*, March 8.

Teotonio, I. (2011) U of T Graduate's Arrest on Terror Charges Alarms Toronto Somalis. *Toronto Star*, March 31.

Turetzky, P. (1998) *Time*. London: Routledge.

Vaughan-Williams, N. (2007) The Shooting of Jean Charles de Menezes: New Border Politics? *Alternatives*, 32(2).

Virilio, P. (2007) *The Original Accident*. Cambridge: Polity Press.

Walzer, M. (2000) *Just and Unjust Wars*. New York: Basic Books.

Weber, C. (2007) Securitizing the Unconscious: The Bush Doctrine of Pre-emption and Minority Report. In Dauphinée, E. and Masters, C. (eds) *The Logics of Biopower and the War on Terror: Living, Dying Surviving*. Basingstoke: Palgrave.

Weber, L. (2007) Policing the Virtual Border: Punitive Pre-emption in Australian Offshore Migration Control. *Social Justice*, 34(2).

Zehfuss, M. (2003) Forget September 11. *Third World Quarterly*, 23(3).

6 While the West sleeps

Deglobed globalization and its consequences

Kevin K. Birth

The representations of time we use to discuss globalization elide the globe and often include unexplored assumptions about simultaneity and speed. When the logic behind the representation of time is explored, one finds a tendency to deny the relevance of Earth's movement to discussions of time. Yet, our bodies are subject to circadian cycles that link us to our position on the globe.

Here I shall discuss several elements of the globalization and administration of Coordinated Universal Time (UTC) in relationship to the emphasis on uniformity in time standards. UTC is essential in the functioning of computer systems. UTC is also kept reconciled with the Earth's unpredictable rotation by means of the leap second policy which inserts (or potentially deletes) one second to keep UTC roughly synchronized with the rotational day. Inserting a leap second is disruptive to computer systems, however. This has led to a proposal to eliminate leap seconds. The push to eliminate leap seconds, and thereby decouple Earth's day and a UTC day, indicates that current time standards privilege uniformity and struggle with the Earth's unpredictable qualities. Yet, this desire for a uniform timescale is not only at odds with the Earth, but consequently also with circadian biology. Daily biological cycles have adapted to the Earth's foibles. So the conflict between uniform timekeeping and the Earth's rotation indicates a larger issue of globalization: we are building a world in which biological cycles and the Earth's motions are suppressed in favor of temporal uniformity. Current policies construct deglobed globalization. The point of this chapter, then, is to show how there is tension between the logic behind the timescale essential to all our electronics and way our bodies have adapted to the Earth's variable rotational cycles. This tension between uniformity and flexibility that is inherent in the relationship of time and globalization has consequences.

In social theory, the focus on clock time often leads to a treatment of the representation of time as equivalent to the logics that create that representation. This confusion prompts Hassan's complaint that discussions about instantaneous communication and time in the media and among many social theorists is ontologically and technologically naïve (Hassan, 2007, pp. 49–50). One such example of this naiveté is Bauman's argument that space and time have "drifted away from each other in human thought and practice" (Bauman, 2000, p. 172). On the contrary, the calculation of the current standard global timescale, UTC, takes into account spatial variables such as the distance time signals travel, the influence of the Doppler effect on those signals, and relativity. In addition, global navigation systems like GPS are actually orbiting clocks with transmitters (see NAVSTAR GPS Joint Program Office, 1996). So representations of time are fundamentally reliant on calculations of space, and determinations of place are fundamentally reliant on calculations of time. Thus, while the thoughts and practices of *some* humans separate space and time, the representations of time and space many people use rely on carefully defined and calculated relationships between space and time. In this case, fetishism not only conceals the process of production, but also conceals how that production entails making decisions and choices that shape users' thought and practice (see Birth, 2012). To study time in relationship to globalization, then, involves the examination of how the representation of time is constitutive of global processes and shaped by human agency (Porter and Stockdale, n.d.).

The cognitive and cultural consequences of clocks

Clocks are cognitive artifacts – items used to extend or enhance cognition (Hutchins, 1999). As Sinha et al. argue, thought about time is "*dependent upon*, and not merely *expressed by*, the employment of symbolic cognitive artefacts" (Sinha et al., 2011, p. 142, emphasis in original). What Dourish says about computers applies to clocks – he writes that computer engineering "is philosophical in the way it represents the world, in the way it creates and manipulates models of reality, of people, and of action" (Dourish, 2001, p. viii).

For instance, in neuroscience there is an effort to identify the biological clock. Throughout the world, there are many rats, hamsters and mice devoted to the determination of the features of this clock, as if evolution had anticipated by 70 million years a device invented in Europe maybe 800 years ago. But there is no clock in the brain (Birth, 2014). There are molecular processes in most organ systems that coordinate rhythms (Hastings et al., 2007),[1] and in mammals these are coordinated by the suprachiasmatic nucleus (SCN) (Dibner et al., 2010). These systems differ from clocks in terms of their adaptability to environmental conditions. Clocks treat time homogeneously, whereas circadian rhythms reflect adaptations to temporal niches (Smarr et al., 2013). The history of horology is one of making clock time uniform, reliable, and unable to be influenced by environmental factors; biology is shaped by evolutionary processes that select for environmental adaptability.

In another example of how clocklike thinking constrains science, Henzi et al. (1992) describe baboons' activity cycles as being, on average, from 09:00 hours to 17:00, or, colloquially, baboons work from 9 to 5. Rather than assuming that 9:00am to 5:00pm is a natural work rhythm for primates, one needs to note the fundamental flaw in the study: baboon circadian rhythms are tied to cycles of daylight, not to clock time. Thus, a more accurate representation of their actual cycle than clock hours would indicate the phase relationship of activity to sunrise and sunset. Yet, I know of only a few field primatology studies that do this (see Erkert, 2008; Erkert et al., 2012; Fernandez-Duque et al., 2010).

In these examples, the lack of understanding of the cognitive tasks for which clocks were designed results in applying clock logic in inappropriate ways. Clocks are very good at measuring duration. In so doing, they represent homogeneous, uniform segments of time. The world is not homogeneous, however. While some questions benefit from knowing exact durations, other challenges are a matter of timing enabling one to act kairotically – i.e. at the right moment. The right moment might be determined by factors that do not recur with regularity and consequently cannot be predicted using measurements of duration. As a result, making the distinction between instants and intervals is critical for the study of some timing behaviors (Wackermann, 2008, pp. 11–12), but clocks entangle these two components of timing (Wackermann, 2008, p. 12; Birth, 2012, pp. 40–41).

Supranational authority over time

The global time system constitutes one element of a new form of empire that creates a global administration and set of laws and conventions that transcend nation-states (Hardt and Negri, 2000). This complex set of practices produces the time to which ordinary clocks are set.

Current time technologies are imperial in their underlying logic. Rather than being tied to a specific place, they seek to be applicable to the entire universe, i.e. the "universal" in Coordinated Universal Time. For those who point to Greenwich Mean Time (GMT) as a global standard that is tied to a particular place – it is not the global standard anymore. Instead, UTC has been the official and preferred term over GMT, and the Royal Observatory in Greenwich has not played the central role in timekeeping since 1957 (Atkinson, 1957). In 1960, UTC replaced GMT as the global standard and the International Bureau of the Hour (BIH) became responsible for managing the global time system.

One could argue that the birth of the modern supranational time system occurred at the International Meridian Conference of 1884. This is one of the earliest postcolonial conferences, in that it consisted of delegates from sovereign states, many of which had gained independence in the 19th century, and it was called by the United States to reach a decision about the placement of the prime meridian and the division of the world into time zones. As Barrows (2011, p. 1) observes, an important issue at the conference was whether nations had sovereignty over time within their borders.

In the second session of this conference, there was a lengthy discussion about whether any resolution passed would be binding on the nations represented there, and there was agreement that it would not be. Yet, because of the nature of such supranational agreements, the outcome of the conference determined the policies followed by most time signal services and map makers, and consequently determined time and navigational practices for nations that officially rejected the outcome of the conference.

The decision of the International Meridian Conference was to place the prime meridian in a smoggy, foggy suburb of London – hardly ideal conditions for the astronomical observation and timekeeping necessary to manage a global timescale. Greenwich claimed the most accurate observations at the time – a claim disputed by the Paris Observatory in France, which led to its vote against the majority opinion. France's apparent compliance despite its public opposition reveals how binding these non-binding decisions can be. In 1911, France set its time to the Paris meridian plus 9 minutes and 25 seconds – a *de facto* synchronization of French time with Greenwich time. At the same time, the French law that did this also stipulated that France would recognize the prime meridian as running through Greenwich once the United Kingdom accepted the metric system (Le Figaro, 1911, p. 1). This happened in 1978.

The International Meridian Conference of 1884 encouraged the creation of supranational institutions to manage the global time system. Global time standards are a critical component of the infrastructure that allows globalization. Terry Quinn, the former director of the International Bureau of Weights and Measures (BIPM) argues that it was globalization of trade that drove the movement toward international standards (Quinn, 2012). Navigation, power grids, telecommunications, and computer technology all rely on robust international time standards, but these standards operate outside the awareness of most clock users. Today, atomic timekeeping standards are managed by the BIPM, and the measurement and prediction of Earth's rotation is the responsibility of the International Earth Rotational and Reference Systems Service (IERS). Together, these agencies now manage several timescales, of which I shall discuss three: UT1, TAI, and UTC.

UT1 represents observed rotational time when taking into account the current position of the poles and varying rotation of Earth. Knowing the current rotational axis allows small corrections to be made in the calculation of mean time (McCarthy & Seidelmann, 2009, p. 226). UT1 is charted by the IERS and reported to the BIPM.

By 1960, the United States and United Kingdom were coordinating their radio time signals based on their observations. This time came to be known as Coordinated Universal Time (UTC). When an atomic standard for the definition of the Système International (SI) second was adopted in 1967 (CGPM, 1967/68, Resolution 1), a problem emerged. Because Earth has a slightly slowing and irregular rotation, 86,400 SI seconds do not equal a rotational day. This problem led to the implementation of adding leap seconds to keep UTC roughly in sync with Earth's rotation.

As a result, a new timescale was devised to be distinct from rotational time – International Atomic Time (TAI). TAI is an atomic timescale consisting of a continuous count of seconds. TAI is not indicated by a master reference clock, but is instead a weighted average of many clocks. This average is calculated by the BIPM using an algorithm known as ALGOS. The algorithm uses a weighted average of times reported by globally distributed atomic clocks to produce TAI (Arias et al., 2011, p. S149). TAI is not related to Earth's rotation – it is a deglobed time. TAI is not distributed, however. It is a timescale that is used by the BIPM to create UTC. The BIPM's official position is that anyone who needs to use TAI can subtract leap seconds from UTC.

UTC then became defined by adjusting TAI by all the leap seconds that have occurred since 1972. This keeps UTC in phase with TAI so that when a TAI second ticks, a UTC second does so simultaneously. UTC also serves to relate TAI to Earth's irregular rotation. When the IERS anticipates a 0.9-second difference between the UTC day and a UT1 day, then a leap second is inserted or deleted – although a second has never actually been deleted.

While the BIPM produces the global timescale and is informed by the IERS about the relationship of UTC to UT1, it is the Radio-communication Sector of the International Telecommunication Union (ITU-R) that determines UTC-related policies, although it does not have any jurisdiction over TAI or UT1. The ITU-R is a United Nations (UN) agency. It is made up of 193 member states that determine global telecommunications policy. The reasons why timescales are administered by the UN telecommunications agency are twofold. First, timescales are distributed by means of telecommunications. Second, it is the timescale that is used for the fundamental determination of all broadcast frequencies – such frequencies defined in terms of number of cycles per second.

The ITU-R is a complicated, postcolonial body. Membership status does not entail agreement with ITU-R decisions. As was the case with the first International Meridian Conference, ITU-R resolutions are only advisory to member states. Even something as seemingly straightforward as the acceptance of UTC as the global timescale requires governmental action within member states (Seago et al., 2011). For instance, while UTC was created in 1960 through the efforts of the United States and United Kingdom, the United States only formally accepted it in 2007 in the America Competes Act (Public Law 110–69, Section 3570). Yet, a consequence of the power of supranational institutions is that they do not need the approval of nation-states to implement policies. The BIPM recognizes ITU-R decisions about UTC regardless of whether ITU-R member states change their laws to reflect those decisions. As a result, an ITU-R-approved change becomes a *de facto* global change even if some nations do not agree with it. Even if a nation passed a law against an ITU-R resolution about time policy, it would have little effect, because of how important BIPM-managed timescales, particularly UTC, are to computer systems, power grids, and telecommunications.

Timestamps

Since most computers cannot process commands in real time, they use UTC timestamps to track changes, manage data, and organize the queue of processes. A computer system's local version of UTC is determined by means of a synchronization protocol with other systems. As a result, computer timestamps are products of at least two supranational systems: 1 the determination of UTC and its distribution; and 2 time synchronization protocols, such as those defined by POSIX, the set of computer systems standards defined and distributed by the Institute of Electrical and Electronic Engineers (IEEE).

For many years, the standard for generating timestamps has been Network Time Protocol (NTP) (Mills, 2011). This is a robust system by means of which networked computers can synchronize with one another to create reliable timestamps. With the growth of the Internet and global connectivity, NTP has become a means of synchronizing computers over large distances.

NTP achieves its robustness through a hierarchy of computer clocks. Computer servers are synchronized through signals exchanged with reliable time sources. At the top of the hierarchy are reference time sources like the United States Naval Observatory and the National Institute of Standards and Technology. At the bottom of the hierarchy are personal devices that synchronize with their networks. A network server should exchange time signal data with several time sources. The times are then run through a set of algorithms to determine the good time sources (the "truechimers") from the bad ones (the "falsetickers"). The time sources that make it through these algorithms are called "survivors." These are then averaged to produce the time to which the local computer is synchronized (Mills, 2011, pp. 37–72).

For some, NTP is not good enough. In 2002, the IEEE generated specifications for Precision Time Protocol (PTP) (IEEE, 2002). This was to be a high-precision time protocol that could achieve precision to levels of around 1 nanosecond. PTP systems have not yet achieved the level of precision in the original IEEE specifications. Still, PTP is much more precise than NTP. Moreover, while PTP systems are not at a 1 nanosecond-level of precision, there is still an option to round one's timestamp to nine digits after the decimal – in effect, to use PTP to generate timestamps as if they were precise to 1 nanosecond.

Two types of institutions use PTP: small-scale institutions that want tight control over very localized automated systems (for example, factories that rely on precision automation), and large-scale multi-billion-dollar global businesses and governmental agencies that can create their own private data links (Seago et al., 2013, p. 47).

Over large networks, PTP is very expensive and difficult to maintain. PTP is so precise that it is important that the routes the time signals travel are symmetrical, i.e. that the outgoing and return trips of the signal are identical so that the signal delay can be accurately calculated. To do this, one sends a signal that is immediately bounced back, and the signal delay is calculated by dividing the time the signal traveled out and back again in half. If it is a long distance, since the Earth is rotating, then the Doppler effect on the signal must also be taken into account. Public networks use routers to avoid creating overused paths, but this causes problems for PTP systems by introducing too much variability in the distances time signals travel. As a result, a PTP system requires the construction of a reliable private data link between all the computers participating in the network. This is expensive and beyond the means of most institutions. To implement a global PTP system for one's network seems to have insurmountable obstacles, yet the global private equity firm IMC Financial Markets is developing this possibility (Estrela and Bonebakker, 2012). Access to timestamp precision becomes a subtle creator of stratification embedded in the production of timestamps.

The technical sophistication of PTP should not obscure the equivocal ontological foundation of many PTP timestamps. PTP relies on the time-steered GPS system which because of steering and factors that affect signal delays from the satellites to receivers is probably accurate to about 10–15 nanoseconds. Yet, PTP can use timestamps that can be set to the precision of 1 nanosecond. At least the NASDAQ uses such timestamps in its data feed TotalView ITCH (NASDAQ, 2014, p. 2), which is used by NASDAQ customers who are able to data-mine stock-trading activities. PTP timestamps are legally recognized, but, as Angel points out, "even clock synchronization at the millisecond level leads to a fundamental uncertainty when latencies are measured in microseconds" (Angel, 2014, p. 275). As the marketing materials for the manufacturers of PTP systems all point out, however, the timestamps generated by their products are acceptable to the US Securities and Exchange Commission (SEC). This implies that merely using a PTP system is sufficient for the SEC – its auditors are not physicists, after all.

The global legal acceptance of PTP despite these issues indicates the extent to which PTP timestamps are cyber speech acts that are increasingly important in managing "big data" and executing market exchanges. Following Austin's classic treatment of these linguistic forms (Austin, 1962), timestamps determine the time of a computer's action by stating the time. As long as a timestamp is within an acceptable range, its veracity is unquestioned. The ability to create fraudulent times is limited, but since the expected ranges for most financial market servers are in microseconds, and PTP functions in nanoseconds, there is the possibility of shifting PTP timestamps within these limits without being noticed. Also, since regulatory bureaucracies lag behind technology, there is a struggle for these bureaucracies to develop rules that address current technological capabilities. The International Organization of Securities Commissions is struggling to define standards in this regard, since right now different nations have different time-protocol standards (OICU-IOSCO, 2013, appendix E).

Global capitalism relies on accurate timestamps. All messages include a timestamp generated by the exchange on which the order is posted, and a collection-point timestamp where the order could be executed. For a traded security, every time its price is adjusted, it receives a new timestamp. Some orders also have expiration times. Together, these collected times constitute what is called "transaction time." Transaction time does not represent a moment in time, but a duration of a piece of data in a database – in this case, an offer in an exchange's trading sequence.

The timestamps that constitute transaction time allow the representation of the sequence of activity pertaining to each order and offered security. This is important because trading volumes are greater than servers can process in real time, so the transaction time system charts the sequence of orders, quotes, and price changes. The rapid global circulation of capital then unfolds in two times – the real time as experienced by people, and a representational time as generated by the sequencing of timestamps. As a recent report by the European Securities and Markets Authority (ESMA) remarked, "[e]xperience shows that, for example, the number of orders received by a trading venue can be very high and in any event, much higher than that of executed transactions" (ESMA, 2014, p. 505). The reason why some orders are not processed is that each order has an effect on price, and once the offering or buying price is outside the parameters of an order, the order is cancelled – the experience that prompted Lewis's book *Flash Boys* (2014; ESMA, 2014). Because quotes change so fast, acting on old quotes is a profound disadvantage, but quotes change so fast that they are not really changing in real time, but in a computed, representational time that is a simulacrum of real time. Until the 2000s, speed of data links was what gave a competitive advantage, but as Angel points out, "speed issues are now bumping up against physical issues. These are not just about computing power, but actual limitations from physics itself" (ESMA, 2014, p. 273).

Since traders cannot violate the laws of physics (Angel, 2014), they are seeking other means to gain a competitive advantage, and PTP seems to be one tool in this effort. It is not clear whether PTP gives an advantage in the position of an order in the queue. It might matter for the ordering of transactions in dark pools (private forums for trading), but few would know this, and anybody who did would recognize that admitting it would be admitting that their dark pool was rigged in somebody's favor. PTP does give a clear and known advantage in data-mining market transactions for purposes of developing time-sensitive algorithms. In current high-frequency markets, timing matters. Angel writes: "that nanosecond just might make the difference between catching a profitable trade and missing it" (Angel, 2014, p. 273). As a result, PTP relying on GPS time signals has become standard among such traders for purposes of tracking orders (Korreng, 2010, p. 361). The ability to use a data feed such as NASDAQ ITCH gives algorithmic traders an immense amount of information about how and when trades occur. Such datasets have been foundational in the emergence of the understanding of market microstructure that has honed the understanding of the influence of timing on price (Dufour and Engle, 2000). This understanding is the foundation of high-frequency trading algorithms. The information gleaned from a temporally precise data feed such as NASDAQ ITCH can be used to adjust the algorithms to anticipate future price changes, and such algorithms not only determine when to sell or make an offer to buy, but also when the offer to buy or the offer to sell expires in its transaction time. Since NASDAQ ITCH uses nanosecond precision, only systems with similar precision can use it to its full advantage – in other words, only PTP systems. Markets are now long past the point when the movement of information began to reach limits imposed by the laws of physics (Angel, 2014). Since the limits of speed have been reached, now it is the ability to take advantage of precision and data mining that give the competitive edge, but only to those who can afford PTP systems.

Leap seconds

Neither NTP nor PTP treats time as related to Earth as a rotating geoid moving through space. Between Earth rotation, human experience, UTC, TAI and PTP-generated simulacra of TAI and UTC, there are actually multiple times that are all mediated by a single representational format – clock time. Current time policy hides these differences through the practice of the leap second. The leap second policy is generated by the need to compromise between the irregularities of the Earth's rotation and the uniformity of atomic timekeeping. The official policy is that when a leap second is necessary, it is announced six months ahead of time by the IERS and is then implemented at midnight at the prime meridian, usually on December 31 or June 30, but it could be the last day of any month. Because leap seconds are not predictable, they cannot be pre-programmed. Instead, when a leap second is announced, systems administrators must install patches in their computer systems to compensate for it.

In an example of the globe disrupting this deglobed timescale, implementing the leap second at midnight at the prime meridian can sometimes mean adding it during rush hour and the beginning of the business day in East Asia, and the end of the business day on the west coast of North America.

During the insertion of a leap second in June 2012, there were several notable systems crashes: Gawker, LinkedIn, Yelp, StumbleUpon, Qantas, Reddit, and Mozilla – all of them except for Gawker involving companies based in the Pacific Rim. The most publicized crash was that of the Qantas airline. It seems that the different servers in the Qantas reservation system started sending packets with timestamps outside the expected range, leading to what is called a livelock – a situation in which two servers are attempting to complete processes related to a connection with each other, but they cannot do so because they keep changing in relationship to each other: "After the leap second happened many sysadmins found that their processors were running, but extremely busy doing operations that were not useful," and that eventually, the solution was simply to reset the clock and date (Allen, 2013, p. 20). Because the reservation system crashed, it was not possible to check in passengers, so the airline ground to a halt for four hours.

This incident has given credence to the argument to eliminate the leap second, and it has been highlighted in a fact sheet prepared by the United States Working Party 7A in preparation for an ITU-R WP 7A meeting (Patten, 2013, p. 9). However, there has been almost no discussion of moving the meridian on which the leap second is implemented, or of changing the implementation day from a calendar date to a date less globally disruptive for a particular leap. The problem is diagnosed not as the position of the meridian or the date on which the leap second is implemented, but as the Earth itself. In this view, globalization requires negating the globe in time standards.

The best known advocates for its elimination have been scientists associated with the United States Department of Defense and its contractors, with the first major publications considering its elimination being co-authored by a number of such scientists (McCarthy and Klepczynski, 1999; Chadsey and McCarthy, 2000; Nelson et al., 2001). Not surprisingly, East Asian nations recognize the problem of leap seconds, with Japanese scientists supporting the American proposal (Koyama et al., 2013). China has offered its own proposal of a leap adjustment once every century (Han, 2014).

Astronomers and those who write software to track anything in the sky are against the proposal because it effectively decouples clock time from Earth's rotation (Seago and Storz, 2003; Finkleman et al., 2011). Some, like the United Kingdom, seem to oppose it because it threatens the importance of the prime meridian, a matter of British heritage, for global timekeeping (Swinford, 2014). Others seem suspicious of any proposal that has the strong support of the US Department of Defense. The global community is not in agreement about how to solve the problem of the disjunction between the timescale needed for globalization of communications and finance, and the behavior of the globe.

The proposal to eliminate the leap second was on the agenda at the 2012 World Radiocommunication Conferences General Assembly. Members of Working Party 7A, the ITU-R committee responsible for timescale policies, polled delegates behind the scenes to discover how they intended to vote. Russell Redman, a Canadian member of WP 7A, reported: "There were a few administrations that stood up and said 'Yes, we must do this!' and a few that stood up and said 'Absolutely not under any circumstances!' but the rest would get very quiet" (Seago et al., 2013, p. 231).

The silence and unwillingness of delegates to vote should not be interpreted as widespread ignorance about the issue. In many cases, it is better understood as informed silence. There is a great deal of technical expertise about computer science, astronomy and metrology around the world, and because of the widespread media attention and duration of the leap second debate, nations' delegates to the ITU-R are probably more informed about the leap second issue than most other ITU-R issues. The unwillingness to take a position probably reflects the postcolonial conundrum the leap second debate creates. In a sometimes-heated debate between representatives from the United States, China, Russia, Germany, France and the United Kingdom, what side should a non-aligned delegate choose? The stakes are high in the sense that the position that wins has great influence over supranational policies and rules governing time. Thus, it is a debate about global political influence rather than just best metrological practice.

Why not ignore the globe in globalization?

Through widespread misapplications, the logic driving precision timekeeping and the desire to eliminate the leap second poses a more serious challenge than the leap second itself. The logic defines time in terms of uniform, homogeneous and unvarying units. This is an effective way to design a computer system, but not a meaningful way to study, say, activity cycles of baboons or the complexity of biological timing systems. In these cases, the use of abstract time measurement without any attention to what triggers timing patterns obscures more than it enlightens. So many cycles and rhythms are tied to place and circumstance, but in contrast, UTC is universal – indeed, it is the official time of all space missions. Biological cycles are about flexibility and adaptation – of being active at the right moment in the right place.

The underlying logic that privileges uniform timekeeping is not limited to time metrology. Because this logic determines clock representations of time, and because of the ubiquity of clocks and their influence on cultural ideas of time, it is what informs most policy decisions involving time, and guides a great deal of human behavior. As Barrows states, the movement toward uniform time removed time and timing from an intrinsic connection to specific labor and replaced labor-based temporalities with a time "applied uniformly to all labor without discrimination" (Barrows, 2011, p. 31). This logic of temporal uniformity is critical to the ideology of the 24/7 society which fosters practices that are "continuous and unbounded" (Crary, 2013, pp. 71–77). It is necessary for the belief that any unit of time is equivalent to any other unit and the erasure of the globe in globalization. This, in turn, is a foundational assumption that the more time devoted to work, the more productive the work will be, as well as the idea that all hours are interchangeable. Such logics elide biology and its effects on behavior.

While there is a tendency in some cultural theory to eschew anything biological as reductive, this serves to empower global power structures that seek to erase the negative health consequences cultivated by current dominant temporalities. Marx argued that the way the body responds to certain exercises of power is not with infinite plasticity. His discussion of the health consequences of poverty in volume one of *Capital* (Marx, 1977, pp. 808–818) reveals that, to Marx, physical suffering was included in the material consequences of capitalism.

The human body's circadian cycles are coordinated by the suprachiasmatic nucleus. The SCN is located at the base of the hypothalamus and contains cells that have pronounced circadian cycles of around 24 hours in duration (Czeisler et al., 1999). These cells are connected with neurons from the optic nerve that have little circadian periodicity, but which respond to light signals. The result is a network of neurons that can be entrained to cycles of daylight and, consequently, keep the body adjusted to those cycles. Since the cycle gets reset by light, it is a system that resets itself daily. As a result, seasonal variability in daylight does not disrupt the circadian rhythm. However, there are limits to the entrainment. These limits seem to be 23.5 hours at the low end and 24.6 hours at the high end (Wright et al., 2001). Beyond these limits, desynchronization occurs. This is when different biological processes are not synchronized with each other, and for many people, the most common experience of this is jet lag. The reason why jet lag is so miserable is that different biological cycles adjust to one's new location at different rates (Waterhouse et al., 1997).

One way to view the SCN is that its periodicity has been selected for by the rotation of Earth and for seasonal variations in daylight. The SCN also adapts circadian biological cycles to the seasonally changing times of sunrise and Earth's rotation caused by the tilt of the Earth interacting with its orbit around the Sun.

The SCN adapts to the Earth's irregularities, but the global timescale has sufficient difficulty with those irregularities so as to seek to reject the relevance of the Earth's movement. Ideologies of globalization privilege temporal uniformity and seek to erase the globe, but they cannot eliminate circadian biology, which remains tied to our rotating geoid. Humans have pronounced circadian rhythms that consist of periods of activity and periods of sleep. These rhythms are critical to our well-being, with many studies having documented serious health risks of chronic sleep deprivation and sleep loss (Knutsson, 2003; Rajaratnam and Arendt, 2001; Cappuccio et al., 2010; Evans and Davidson, 2013), and evidence that work conditions contribute to class-based health disparities (Nachreiner, 2004; Burgard and Lin, 2013). These risks include much higher incidence of diabetes, cardiovascular disease, obesity, and cancer. They also include impairment of cognitive function – particularly the evaluation of risk and memory functions (Goel et al, 2009; Womack et al., 2013). To treat all hours of the day as equivalent is biologically unsound, and yet the 24/7 society tends to do this, despite horrific accidents and outcomes, such as the nuclear accident at Chernobyl which has been attributed to impaired cognitive function due to fatigue (Moore-Ede, 1993, pp. 5–6, 108–109). As Moore-Ede (1993, p. 5) puts it, we are "not built for a world we have designed." These consequences have been demonstrated in the frequency of accidents and mistakes made by those who are sleep starved. If one thinks on a grander scale, and looks at the professions in which one is most likely to experience chronic sleep loss, then the ubiquity of the problems generated by the 24/7 society and the assumption of temporal uniformity begin to come into view.

According to Luckhaupt et al. (2010), the most sleep-starved occupation in the United States is that of corporate managers and executives. Those who manage the banking and financial system are among those most affected by sleep loss, which means that they are a population particularly susceptible to memory impairment and a diminished ability to evaluate risk. The cause of the recent global economic recession was linked to inaccurate evaluations of financial risks.

Next on the list are those involved in transportation and warehousing. A week does not go by in New York City without the report of a horrific accident caused by a sleepy driver. In 1990, the National Transportation Safety Board determined that the most common cause of heavy-truck crashes in which the driver was killed was fatigue (National Transportation Safety Board, 1990).

In the United States, even in the least sleep-deprived industry, agriculture, the statistics indicate that 23 percent of workers are sleep starved.

This sort of sleep starvation belies the concept of time-space compression. Harvey argues that the "annihilation" of space is a consequence of time-space compression (Harvey, 1989, p. 248; Harvey, 1990, p. 425). Space is only annihilated for the privileged, however. Even then, it is only partially annihilated. Our circadian biology retains the cycles of the last place in which we resided. We are biologically tied to our place on the globe – a facet of globalization that is often neglected. What appears as time-space compression is one facet of what Sharma describes as the "uneven cultural politics of time" (Sharma, 2014, p. 7). By focusing on lived time, she describes systems of stratification associated with time-space compression and speed. The services to provide slow food or slow living, taxi transportation, or solace to the jet lagged, or 24/7 customer support involve disrupting and exploiting the temporal rhythms of the service providers. Those people warp their schedules to serve privileged people whose bodies retain the cycles of their previous locations even as they move about the globe. These types of service providers are well aware of their location in relationship to cycles of daylight, and are very conscious of the fatigue from which they suffer due to working irregular shifts in order to accommodate privileged people from other locations who want to think, like Thomas Friedman (2005), that "the world is flat."

Even among those to whom time-space compression applies, they still live on a globe. This means that some are privileged to have their work and leisure events occur when it is convenient, whereas others have to warp their work and leisure schedules to suit interests half a world away, thereby cultivating sleep loss. In an example I have discussed elsewhere, in the 2004 Olympic Games held in Athens, the marathon was scheduled in the heat of the day in order to maximize the television audience, rather than during the morning hours when conditions are best for the health and welfare of the runners (Birth, 2007, p. 224; Birth, 2012, pp. 145–146). When one seeks customer support during the day only to get someone seemingly incompetent on the other side of the globe, one should remember that during the early morning hours, one is least cognitively capable.

Misleadingly deglobed globalization

The lack of attention to the biological consequences of time and timing as well as the tendency to ignore the globe in globalization has promoted a discussion of time primarily in terms of its commodification and speed. Many in the media and among those who create public policy treat computer technology as magical without its inner workings being explored, and are mesmerized by the simple representation of time without understanding time metrology. Even when time metrology is discussed, the reports are often hastily written and full of errors. Contrary to popular imagination, there are no instantaneous communications over networks, and computers do not run in real time but rely on timestamps to order tasks. When computer technology is examined, then one realizes that engineers reached the limits of speed in the 1990s, and it has been since then that the emphasis has been on increasing precision. In discussing time and globalization the consequences of living on our rotating geoid and the effects of our actions and policies on health and safety must be addressed. People die because of the assumption of the 24/7 temporal uniformity on which commodified time is based. It has real biological consequences. In our theories, analyses and policies, globalization may be deglobed, but our bodies do not allow such facile denial of the planet on which we live.

These are consequences we will all, at some point, experience, whether it is due to a mistake made by a medical professional who has worked too many hours; or a traffic jam caused by an accident involving a sleep-deprived driver; or ineffective customer support from a representative who, because they are 12 time zones away, is working in the middle of the night; or a large loss in one's retirement account because of financial analysts who rather than accurately evaluating risk, doubled down on their risky assumptions; or even our own death.

Acknowledgments

An earlier version of this chapter was presented at the Temporal Design Workshop in Edinburgh in May 2014. I thank the participants in that workshop for their comments. I also thank Ken Seidelmann, Rob Seaman, Steve Allen, and Jim Angel, Michelle Bastian and the anonymous reviewers for their general comments and for their attempts to ensure that my representation of current timescale dissemination and market practices are correct. Any errors that remain are my responsibility.

Note

1 Rather than cite the voluminous literature in chronobiology, I cite articles that allow a strategic entry into this prolific field.

References

Allen, S.L. (2013) Planes Will Crash! Things that Leap Seconds Didn't and Did, Cause. In Seago, J.H., Seaman, R.L., Seidelmann, P.K. and Allen, S.L. (eds) *Requirements for UTC and Civil Timekeeping on Earth* (pp. 15–24), American Astronautical Society Science and Technology Series, 115.

Angel, J. (2014) When Finance Meets Physics: The Impact of the Speed of Light on Financial Markets and their Regulation. *Financial Review*, 49, pp. 271–281.

Arias, E.F., Panfilo, G. and Petit, G. (2011) Timescales at the BIPM. *Metrologia*, 48, pp. S145–S153.

Atkinson, R. (1957) The Longitude of Herstmonceux. *The Observatory*, 77, pp. 31–32.

Austin, J.L. (1962) *How to Do Things with Words*. Cambridge, MA: Harvard University Press.

Barrows, A. (2011) *The Cosmic Time of Empire: Modern Britain and World Literature*. Berkeley, CA: University of California Press.

Bauman, Z. (2000) Time and Space Reunited. *Time and Society*, 9, pp. 171–185.

Bede (1999) *The Reckoning of Time*. (Faith Wallis, trans.). Liverpool: Liverpool University Press.

Birth, K.K. (2007) Time and the Biological Consequences of Globalization. *Current Anthropology*, 48, pp. 215–236.

Birth, K.K. (2012) *Objects of Time: How Things Shape Temporality*. New York: Palgrave Macmillan.

Birth, K.K. (2014) Non-Clocklike Features of Psychological Timing and Alternatives to the Clock Metaphor. *Timing and Time Perception*, 2(3), pp. 312–324.

Burgard, S.A. and Lin, K.Y. (2013) Bad Jobs, Bad Health? How Work and Working Conditions Contribute to Health Disparities. *American Behavioral Scientist*, 57, pp. 1105–1127.

Cappuccio, F.P., Miller, M.A. and Lockley, S.W. (eds) (2010) *Sleep, Health and Society: From Aetiology to Public Health*. Oxford: Oxford University Press.

CGPM (1967/8) Resolution 1: SI Unit of Time (second). The 13th Conférence Générale des Poids et Mesures (CGPM). www.bipm.org/en/CGPM/db/13/1/.

Chadsey, H. and McCarthy, D. (2000) Relating time to the Earth's variable rotation. Proc. 32nd Annual Precise Time and Time Interval (PTTI) Systems and Applications Meeting (pp. 28–30). Reston, VA.

Cont, R., Kukanov, A. and Stoikov, S. (2011) The Price Impact of Order Book Events. arxiv.org/pdf/1011.6402.

Crary, J. (2013) *24/7: Late Capitalism and the Ends of Sleep*. London: Verso.

Czeisler, C.A. (2010) Ethical Considerations for the Scheduling of Work in Continuous Operations: Physicians in Training as a Case Study. In Francesco, P., Cappuccio, M., Miller, A. and Lockley, S.W. (eds) *Sleep, Health, and Society: From Aetiology to Public Health* (pp. 435–456). Oxford: Oxford University Press.

Czeisler, C.A., Duffy, J.F., Shanahan, T.L., Brown, E.N., Mitchell, J.F., Rimmer, D.W., … Kronauer, R.E. (1999) Stability, Precision, and Near-24-h Period of the Human Circadian Pacemaker. *Science*, 284, pp. 2177–2181.

Dibner, C.A., Schibler, U. and Albrecht, U. (2010) The Mammalian Circadian Timing System: Organization and Coordination of Central and Peripheral Clocks. *Annual Review of Physiology*, 72, pp. 517–549.

Dourish, P. (2001) *Where the Action is: The Foundations of Embodied Interaction*. Cambridge, MA: MIT Press.

Dufour, A. and Engle, R.F. (2000) Time and the Price Impact of a Trade. *Journal of Finance*, 55, pp. 2467–2498.

Erkert, H.G. (2008) Diurnality and Nocturnality in Nonhuman Primates: Comparative Chronobiological Studies in Laboratory and Nature. *Biological Rhythm Research*, 39, pp. 229–267.

Erkert, H.G., Fernandez-Duque, E., Rotundo, M. and Schneider, A. (2012) Seasonal Variation of Temporal Niche in Wild Owl Monkeys (Aotus azarai azarai) of the Argentinean Chaco: a Matter of Masking? *Chronobiology International*, 29, pp. 702–714.

European Securities Markets Authority (ESMA) (2014) *Discussion Paper MiFID II/ MiFIR.*

Estrela, P.V. and Bonebakker, L. (2012) Challenges deploying PTPv2 in a Global Financial Company. 2012 International IEEE Symposium on Precision Clock Synchronization for Measurement Control and Communication.

Evans, J. and Davidson, A. (2013) Health Consequences of Circadian Disruption in Humans and Animal Models. *Progress in Molecular Biology and Translational Science*, 119, pp. 282–323. DOI: 10.1016/B978-0-12-396971-2.00010-00015.

Fernandez-Duque, E., de la Iglesia, H.O. and Erkert, H.G. (2010) Moonstruck Primates: Owl Monkeys (Aotus) Need Moonlight for Nocturnal Activity in their Natural Environment. *PLoS One*, 5, e33981.

Finkleman, D., Allen, S., Seago, J.H., Seaman, R. and Seidelmann, P.K. (2011) The Future of the Leap Second. *American Scientist*, 99, pp. 312–319.

Friedman, T.L. (2005) *The World is Flat: A Brief History of the Twenty-First Century.* New York: Farrar, Straus and Giroux.

Goel, N., Rao, H., Durmer, J.S. and Dinges, D.F. (2009) Neurocognitive Consequences of Sleep Deprivation. *Seminars in Neurology*, 29, pp. 320–339.

Han, C. (2014) Space Odyssey: Time-scales and Global Navigation Satellite Systems. *ITU News.* Presented January 2013. itunews.itu.int/En/4273-Space-odyssey-BRTim e-scales-and-global-navigation-satellite-systems.note.aspx.

Hardt, M. and Negri, A. (2000) *Empire.* Cambridge, MA: Harvard University Press.

Harvey, D. (1989) *The Condition of Postmodernity.* Oxford: Blackwell.

Harvey, D. (1990) Between Space and Time: Reflections on the Geographical Imagination. *Annals of the Association of American Geographers*, 80, pp. 418–434.

Hassan, R. (2007) Network Time. In Hassan, R. and Purser, R.E. (eds) *24/7: Time and Temporality in the Network Society* (pp. 37–61). Palo Alto, CA: Stanford University Press.

Hastings, M., O'Neill, J.S. and Maywood, E. (2007) Circadian Clocks: Regulators of Endocrine and Metabolic Rhythms. *Journal of Endocrinology*, 195, pp. 187–198.

Henzi, S.P., Byrne, R.W. and Whiten, A. (1992) Patterns of Movement by Baboons in the Drakensberg Mountains. *International Journal of Primatology*, 13(6), pp. 601–629.

Hewlett, S.A. and Luce, C.B. (2006) Extreme Jobs: The Dangerous Allure of the 70-Hour Workweek. *Harvard Business Review*, 84, pp. 49–59.

Ho, K. (2009) *Liquidated: An Ethnography of Wall Street.* Durham, NC: Duke University Press.

Hutchins, E. (1999) Cognitive Artifacts. In Wilson, R.A. and Keil, F.C. (eds) *MIT Encyclopedia of the Cognitive Sciences* (pp. 126–127). Cambridge, MA: Massachusetts Institute of Technology Press.

IEEE (2002) *IEEE 1588–2002: IEEE Standard for a Precision Clock Synchronization Protocol for Networked Measurement and Control Systems.*

International Meridian Conference (1884) *International Conference Held at Washington for the Purpose of Fixing a Prime Meridian and a Universal Day: Protocols of the Proceedings.* Washington, DC: Gibson Bros.

Knutsson, A. (2003) Health Disorders of Shift Workers. *Occupational Medicine*, 53, pp. 103–108.

Korreng, M.D. (2010) UTC Time Transfer for High Latency Trading Using IS-95 CDMA Base Station Transmissions and IEEE-1588 Precision Time Protocol. In Breakiron, L. (ed.), *42nd Annual Precise Time and Time Interval (PTTI) Meeting* (pp. 359–368). Manassas, VA: Institute of Navigation.

Koyama, Y., Iwama, T., Ito, H., Hanado, Y. and Hsokawa, M. (2013) Experiences of Leap Second Adjustment Operations and Questionnaires in Japan. In Seago, J.H., Seaman, R.L., Seidelmann, P.K. and Allen, S.L. (eds) *Requirements for UTC and Civil Timekeeping on Earth* (pp. 27–32). American Astronautical Society Science and Technology Series, 115.

Le Figaro (1911). L'Heure Anglais, March 11, 1.

Lewis, M. (2014) *Flash Boys: A Wall Street Revolt*. New York: Norton.

Luckhaupt, S.E., Tak, S. and Calvert, G. (2010) Prevalence of Short Sleep Duration by Industry and Occupation in the National Health Interview Survey. *Sleep*, 33, pp. 149–159.

Marx, K. (1977) *Capital, v. 1*. (B. Fowkes, trans.). New York: Vintage.

McCarthy, D. and Klepczynski, W. (1999) GPS and Leap Seconds: Time to Change? *GPS World*, November, 50–54.

McCarthy, D. and Seidelmann, P.K. (2009) *Time: From Earth Rotation to Atomic Physics*. Weinheim: Wiley-VCH.

Mills, D. (2011) *Computer Network Time Synchronization*, 2nd edn. Boca Raton, FL: CRC Press.

Moore-Ede, M. (1993) *The Twenty Four Hour Society*. Reading, MA: Addison-Wesley.

Nachreiner, F. (2004) Diversity and Equity: Dealing with Biological and Social Differences. *Revista Saúde Pública*, 38 (supplement), pp. 80–85.

NASDAQ (2014) NASDAQ TotalView-ITCH 4.1. www.nasdaqtrader.com/content/technicalsupport/specifications/dataproducts/NQTV-ITCH-V4_1.pdf.

National Transportation Safety Board (US) (1990) *Safety Study: Fatigue, Alcohol, Other Drugs, and Medical Factors in Fatal-to-the Driver Heavy Truck Crashes*. Washington, DC: National Transportation Safety Board.

NAVSTAR GPS Joint Program Office (1996) *NAVSTAR GPS User Equipment Introduction*. www.navcen.uscg.gov/pubs/gps/gpsuser/gpsuser.pdf.

Nelson, R.A., McCarthy, D.D., Malys, S., Jevine, J., Guinot, B., Fliegel, H.F., Beard, R.L. and Bartholmew, T.R. (2001) The Leap Second: Its History and Possible Future. *Metrologia*, 38, pp. 509–529.

OICU-IOSCU (2013) *Technological Challenges to Effective Market Surveillance Issues and Regulatory Tools, Final Report*. The Board of the International Organization of Securities Commissions.

Patten, B.M. (2013) *Fact Sheet: Proposed Modifications to Annex 3 of Document 7a/28, Working Party 7A Chairman's Report – Preliminary Draft New Report ITU-R TF [UTC BACKGROUND] USWP 7A*.

Porter, T. and Stockdale, L. (n.d.) *The Strategic Manipulation of Transnational Temporalities*.

Quinn, T. (2012) *From Artefact to Atoms*. Oxford: Oxford University Press.

Rajaratnam, S. and Arendt, J. (2001) Health in a 24-h Society. *The Lancet*, 358, pp. 999–1005.

Rotter, G.S. (1969) Clock-speed as an Independent Variable in Psychol Res. *The Journal of General Psychology*, 81, pp. 45–52.

Seago, J.H., Seaman, R.L., Seidelmann, P.K. and Allen, S.L. (eds) (2013) *Requirements for UTC and Civil Timekeeping on Earth*. American Astronautical Society, Science and Technology Series, 115.

Seago, J.H., Seidelmann, P.K. and Allen, S. (2011) Legislative Specifications for Coordinating with Universal Time. In Seago, J.H., Seaman, R.L. and Allen, S.L. (eds) *Decoupling Civil Timekeeping From Earth Rotation* (pp. 29–48).

Seago, J. and Storz, M.F. (2003) UTC Redefinition and Space and Satellite-Tracking Systems. Proceedings of the ITU-R SRG Colloquium on the UTC Timescale, IEN Galileo Ferraris, Torino, Italy, pp. 28–29.

Sharma, S. (2014) *In the Meantime: Temporality and Cultural Politics.* Durham, NC: Duke University Press.

Sinha, C., da Silva Sinha, V., Zinken, J. and Sampaio, W. (2011) When Time is not Space: The Social and Linguistic Construction of Time Intervals and Temporal Event Relations in an Amazonian Culture. *Language and Cognition*, 3, pp. 13–169.

Smarr, B.L., Schwartz, M.D., Wotus, C. and de la Iglesia, H.O. (2013) Re-examining "Temporal Niche." *Integrative and Comparative Biology*, 53, pp. 165–174.

Swinford, S. (2014). Greenwich Mean Time could drift to the US, minister warns. *Daily Telegraph*, May 14. www.telegraph.co.uk/news/politics/10831974/Greenwich-Mean-Time-could-drift-to-the-US-minister-warns.html.

Wackermann, J. (2008) Measure of Time: A Meeting Point of Psychophysics and Fundamental Physics. *Mind and Matter*, 6, pp. 9–50.

Waterhouse, J., Reilly, T. and Atkinson, G. (1997) Jet-lag. *Lancet*, 350(9091), pp. 1611 1616.

Womack, S.D., Hook, J.N., Reyna, S.H. and Ramos, M. (2013) Sleep Loss and Risk-Taking Behavior: A Review of the Literature. *Behavioral Sleep Medicine*, 11, pp. 343–359.

Wright, K.P., Jr., Hughes, R.J., Kronauer, R.E., Derk-Jan, D. and Czeisler, C.A. (2001) Intrinsic near-24-h Pacemaker Period Determines Limits of Circadian Entrainment to a Weak Synchronizer in Humans. *Proceedings of the National Academy of Sciences of the United States of America*, 98, pp. 14027–14032.

7 Accelerated contagion

Understanding the relationships among globalization, time, and disease

Yanqiu Rachel Zhou and William D. Coleman

Introduction

The transmission in 2003 of Severe Acute Respiratory Syndrome (SARS) around the world "at the speed of a jet airplane" (Health Canada, 2003, p. 23) raises questions about the complex intersections of globalization, time, and diseases. Contemporary globalization processes have deterritorialized the world through global financial markets and rapid expansion in modes of travel in all parts of the sphere (Aaltola, 2012). Through the emergence of global networks of transportation and communication that link countries and cities in new, more integrated ways, increasing transnational contacts, travel, and integration have given rise to new health threats. Not surprisingly, international and global health organizations have framed the new challenges in terms of security. Primarily they use spatiality-based measures (e.g. border control); this framing, however, overlooks the important changes in the nature and experience of time that have accompanied economic, political, cultural, and social globalizing processes. These processes have led to the "speeding up" of time, to new forms of time, and to changes in the relationship between time and space/place, as well as between different temporalities. These changes, in turn, produce such phenomena as "time-space compression," "timeless time," "spaces of flow," "simultaneity" and "multi-temporality" (Castells, 2009; Harvey, 1990; Rosa, 2009; Scholte, 2005).

In recent years the emerging infectious diseases (EIDs) such as SARS, avian flu, H1N1, Ebola, and Middle East Respiratory Syndrome (MERS) testify to the difficulties globalizing processes present for public health surveillance and interventions. The increased global movements of people and microbes have generated temporal uncertainties when it comes to the speed of transmission and the pressure to respond quickly to disease outbreaks. The fast trans-border transmission of infectious agents arises from steep increases in the volume, frequency, and scope of international travels of humans, animals and plants which spread pathogens. These difficulties generate questions about traditional state-centric, often spatiality-based, approaches to disease control and health governance.

Technological advances since the 1970s have accelerated contemporary globalizing processes which, in turn, have changed the speed of the circulation of pathogens. In response, actors at all levels of global public health are pressed to speed up their responses. Differences in the experiences of, and capacities for, "speeding up" of time have created profound new challenges to both domestic responses and global collaboration to control the spread of infectious diseases. Despite proliferation of globalizing linkages among national economies, the continued legitimacy and importance of state borders also pose challenges to addressing trans-border phenomena related to health and disease.

Viewing it as a disease of speed, this article examines SARS as a case of EIDs in the context of contemporary globalization. Primarily drawing on theories of "acceleration" and the "global cities network," we explore the challenges and possibilities of global health exemplified by this disease in a globally interconnected and, increasingly, interdependent world. This theoretical framework allows us to understand time as both a standardized overarching system (e.g. "universal" clock time) and as a multi-dimensional construct associated with technology, place, politics, and history. It also assists in studying the intersectional relationships among globalization, time and disease. Guided by these theories, we present our case study in two parts. First we review the 2003 SARS crisis from a temporal perspective – in particular, its accelerated spread through the global cities network – and second the rapid, yet temporally contested, responses by some major affected countries to the simultaneity of the crisis. In the section that follows, we examine how the response to SARS at a global level suggests a promising new approach to managing temporal challenges posed by EIDs. We contend that the accelerated transmission of SARS has exposed the limitations of traditional spatiality-based approaches to infectious diseases, disease control and health governance. When the advances in information and communication technologies (ICTs) in recent decades have accelerated the diffusion of pathogens, actors at all levels of global public health are pressed to keep up with the new temporalities.

Conceptualizing the relationships among globalization, time, and disease

In their book *Timespace: Geographies of Temporality*, May and Thrift view the late 20th century as one of the historical periods when society witnessed "a significant acceleration in the pace of life concomitant with a dissolution or collapse of traditional spatial co-ordinates (changes usually expressed via some kind of discourse on *speed* – or space divided by time)" (2001, p. 7). Viewing modern societies as acceleration societies, Rosa (2013) also depicts three main, mutually reinforcing types of acceleration that constitute a "circle of acceleration." That is, "*technological acceleration* tends to increase the pace of *social change*, which in turn unavoidably increases the experienced *pace of life*, which then induces an ongoing demand for technical acceleration in the hopes of saving time, and so on back around the circle" (Rosa, 2013, p. xx, emphasis added). In addition, disease transmission is accelerating due to the pace of ecological and environmental change that brings about new animal-human interfaces, and increasing human mobility (Bashford, 2006; Weiss & McLean, 2005).

In discussing the relationship between globalization and health, Lee (2003) argues that three types of boundary – spatial, temporal, and cognitive – have eroded and been redefined, because globalization processes have changed the nature of human interaction by intensifying the interactions across these boundaries that "have hitherto separated individuals and population groups from each other" (p. 21). The intensification and diversification of human contacts have generated two forms of spatial changes: 1) a redefinition of existing territorially based geographies; and 2) an increasing degree of social interaction that is detached from territorial spaces, such as e-commerce (Lee, 2003). Viewing EIDs, for example, as a security issue has motivated nation-states to reinforce national borders on one hand. On the other hand, the deterritorialized potential of EIDs also calls for collaboration that goes beyond national and regional levels of border control (Aaltola, 2012; Ingram, 2005; WHO, 2006). Enabled by technological advances, geographical deterritorialization in the context of accelerated human interaction leads to the spread of changes in lifestyles (including related health conditions) and the quick movement of infectious agents across geographies (Lee, 2003). Spatial and temporal changes also result in alterations to the creation and exchange of information, ideas, beliefs, norms, and other thought processes. For example, sharing of knowledge globally through ICTs can lead to more rapid adjustments in knowledge and practices in health interventions and governance (Lee, 2003). During the SARS crisis, the World Health Organization's (WHO) travel advisory targeted individual travelers rather than nation-states, and thereby facilitated faster dissemination of information (WHO, 2006).

Coining the concept of "time-space compression," Harvey links acceleration with the history of capitalism, which "has been characterized by speed-up in the pace of life" (1990, p. 240). He points to changes in time arising from large corporations moving away from vertical structures to outsourcing and sub-contracting, which, in turn, quickened the production and assembling of goods. When the spatial barriers are overcome by technologies, in effect, the distances between places – measured by, for example, travel time or cost – are reduced or even annihilated, and time horizons are "shortened to the point where the present is all there is" (Harvey, 1990, p. 240). Consequently, at a global level, "space appears to shrink to a 'global village' of telecommunications and a 'spaceship earth' of economic and ecological interdependences" (Harvey, 1990, p. 240). On a social level, however, people have to learn how to cope with "an overwhelming sense of *compression* of our spatial and temporal worlds" (Harvey, 1990, p. 240, author's emphasis). The rapidity of the global spread of SARS through international air travel speaks to the relevance of the concept of time-space compression.

Historically, time was defined as the sequencing of practices, in such terms as "biological time," "clock time" (the industrial age), and "social time"

(Castells, 2009, pp. 34–35). In contrast, in the present era, as time "accelerates," sequencing is lost. Novel ICTs compress time to the point that the sequences of social practices (past, present, future) blur. The instantaneity of information transmission means that all events appear to be simultaneous in digital communication. In these respects, time becomes "timeless" (Castells, 2009, p. 35). More and more parts of the world experience the same phenomena as "no time" and "at the same time" (Scholte, 2005, p. 62), whether these be a banking crisis, extreme weather, or an EID. Under these circumstances, individuals' actions also illustrate a growth in "the scope and depth of consciousness of the world as a single place" (Scholte, 2005, p. 267), including "a place" that might fight the same EID anywhere in the world at the same time.

Taking Harvey's observations further, Castells (2009) develops the concept of "spaces of flows" to capture the new spatial reality: "the technological and organizational possibility of practicing simultaneity without continuity" (Castells, 2009, p. 34). Spaces of flows are made of the articulation between three elements: "the places where activities (and people enacting them) are located; the material communication networks linking these activities; and the content and geometry of the flows of information that perform the activities in terms of function and meaning" (Castells, 2009, p. 34). Built as they are on "timeless time," such social spaces also mean "the possibility of asynchronous (not synchronous) interactions in chosen time" (Castells, 2009, p. 34). The spaces of flows, working through the social form of the *network*, are increasingly used to respond quickly to the accelerated spread of infectious diseases.

The network geographies of spaces of flows and of the experience of timeless time map onto, in turn, the "global cities network." Building on Sassen's research on "global cities" that was limited to a few key nodes like New York, London and Tokyo (Sassen, 2001), Taylor (2004) argues that a growing number of cities, including former "Third World" cities like Hong Kong, Shanghai, Singapore and Taipei, have become the central nodes of the globalizing processes, as he explains:

> The experience of cyberspace is not essentially hierarchical; it operates as innumerable networks, albeit across an uneven globalization. In this sense, then, all cities are global: they operate in a contemporary space of flows that enables them to have a global reach when circumstances require such connections.
>
> (Taylor, 2004, p. 43)

This specification of new social structures built around global cities adds to our analysis, because these same horizontal city-to-city networks become the primary conduits for the rapid global spread of infectious diseases. In their study of emerging infections of SARS in global cities, Ali and Keil (2008, p. 5) argue that the global cities network could also potentially "serve as a network for disease transmission," given the intensified flow of people in these ever-dynamic hubs.

While space can be "annihilated" by time, the role of place as a geographical site or physical space in constituting temporalities has remained important. According to Sassen (2000), in a global city we can see the coexistence of an old, or collapsing, temporality (the time of the nation-state as an historical institution) and a new temporality (the time of economic globalization). The intersection of these different expressions of time generates new dynamics and opportunities that drive economic, political, and social globalizations, and "can be thought of as partly de-nationalized temporalities" (Sassen, 2000, p. 20). For example, global networks of finance, transnational corporations, and international organizations routinely impose their temporal priorities in local contexts, in which there exist not only different temporalities but also different capacities to respond to globally desirable temporal frameworks. Given that different groups and segments of society have different capacities to synchronize with global processes, these changes result in an increasing multi-temporality (Rosa, 2009):

> This desynchronization entails an increasing "simultaneity of the non-simultaneous": high-tech and stone-age methods of warfare, transport, or communication persist side by side, not only between different countries, but even within the same society, and fast and slow paces of life can be observed on one and the same street.
>
> (Rosa, 2009, pp. 103–104)

In short, uneven globalizing processes have complicated the temporal relationships – in such forms as temporal inequality and disjuncture – among people and societies. In the next section, we explore the impacts of such relationships on the experiences of SARS that demanded fast, simultaneous action across geographies.

The temporalities of contagion: the case of SARS

Accelerated transmission: the global spread of SARS

Emerging in southern China in November 2002, SARS, a viral respiratory disease caused by a novel coronavirus, encompassed the globe, to varying degrees, within weeks. November 16 marked the first retroactively identified case, in Foshan city in China. In less than two months it broke out in southern China, with similar outbreaks about one month later in other parts of China and Asia (e.g. Hong Kong, Vietnam, and Singapore), and in Vancouver and Toronto in Canada (WHO, 2006; Health Canada, 2003). The speed with which SARS spread was explicitly linked with "the age of globalization" in the WHO's 307-page report titled "SARS: How a global epidemic was stopped." Specifically, mass, rapid, international travel enabled the transformation of SARS from a local outbreak into a global pandemic; in the absence of accelerated transworld travel, "it would probably have remained a localized problem, with few consequences for global health" (WHO, 2006, p. viii).

As one of the "technologies of speed," jet passenger aircrafts were instrumental in the rapid spread of SARS across countries and continents. Fast air travel means that it takes only hours for a SARS-infected individual, either symptomatic or asymptomatic, to move from one place to another. Such a move becomes even more dangerous when the time used for travel is much less than the 2–10-day incubation and infection period of the SARS Co-Virus (Ali, 2008, p. 244). During the three-hour travel time of Flight CA112 from Hong Kong to Beijing widespread SARS transmission occurred. In turn, the passing on of the infection is sufficiently rapid that the carrier of the disease to Beijing easily infected travelers both remaining in Beijing and flying on to other cities, such as Taipei, Singapore, and Bangkok (WHO, 2006). Furthermore, the aircraft itself constitutes a distinct environment: one in which passengers are vulnerable to airborne pathogens due to its confined space, little physical mobility, and shared, re-circulated air (Mangili and Gendreau, 2005; Budd et al., 2009; Ali and Keil, 2006). In short, the high degree of mobility of infected, asymptomatic individuals through the networks of international airlines increased the risk of widespread transmission, despite control measures at airports (Mangili and Gendreau, 2005; WHO, 2006). Although the specific patterns of in-flight transmission are yet to be determined, SARS exemplifies the real potential for aircrafts to function as disease "amplifiers" for "borderless" transmission (Mangili and Gendreau, 2005).

In addition, the global hub-and-spoke networks of air transportation overlap directly with the network of "global cities" (Sassen, 2001; Taylor, 2004, 2013). Providing the infrastructure for global capitalism, the frequent, very fast movement of business and support experts increases the probability of further infection. For example, 78-year-old Ms KSC returned to Toronto, a global city, from Hong Kong, also a global city, on February 23, and passed the virus on to four members of her extended family before she died; these infections, in turn, sparked the Toronto outbreak (Health Canada, 2003; WHO, 2006). Mr LSK, who acquired SARS in Hong Kong, transmitted it to at least 22 passengers and two crew members (including residents from Hong Kong, Taiwan, Singapore, and a passenger who flew on to Bangkok) on flight CA112 on March 15, 2003; and at least 59 people were infected after his arrival in Beijing (WHO, 2006). As a result, *simultaneous*, multi-directional transmission took place in these cities.

Unlike the "spatially contagious diffusion" (spread from major regional epicenters to smaller places) of HIV at an international level, the global spread of SARS occurred between global cities moving along their global economic connections (Ali and Keil, 2006). Given Hong Kong's status as a top "global city" (Alpha plus) that is more integrated with the global economy than any other city except London and New York (GaWC, 2012), it is not surprising that the city became "an important interchange site" for the global spread of SARS (Ali and Keil, 2006, p. 500). It is also no a coincidence that accelerated SARS transmission did not take place until the index case of the Metropole Hotel outbreak arrived in the global city of Hong Kong from Guangdong

Province on February 21, 2003. Although the virus had stayed in Guangdong since its first case in mid-November 2002, the diffusion of the disease was accelerated immensely once it arrived in the global cities network. After arriving in Hong Kong, it only took two days to reach Toronto (on February 24, 2003), and three more weeks to arrive massively in Beijing (on March 15, 2003) (WHO, 2006). Although the global media focused on Asia or China as the cause of the pandemic, they overlooked the temporal significance of Hong Kong as a lead node in the global cities network. This network also includes other SARS-affected cities that are classified as first-tier (Alpha category) global cities, such as Beijing (Alpha plus), Singapore (Alpha plus), Toronto (Alpha), Bangkok (Alpha minus) and Taipei (Alpha minus). This ranking means that their contributions to supporting the global economy are of the highest degree (GaWC, 2012). Had the infection not been identified and contained by April 2003, the global cities network could have facilitated even wider disease outbreaks given its considerable potential for accelerating the pandemic.

The fluidity of human flows in the global cities network accelerated the rapid spread of SARS. Indeed, most of the "index cases," or "super-spreaders," who passed the virus on to many others, and thus sparked local outbreaks, were frequent travelers. The index case that started the global spread of SARS was a professor of medicine who went to Hong Kong for a relative's wedding from Guangzhou (China), where he was treating SARS-infected patients. The Hanoi index case was a New York businessman who had traveled to China and Hong Kong before arriving in Vietnam. The source case on flight SQ 25 (from New York to Singapore) was a doctor who treated SARS cases in Singapore, and then attended a medical conference in New York. The source cases on flight TG614 (from Bangkok to Beijing) were two Chinese officials who became infected on flight CA112, and passed the virus on to a Finnish official of the International Labour Organization, who had traveled in Europe and Bangkok before going to Beijing (WHO, 2006).

The rapid expansion and intensification of global air traffic flows along the global cities network have been instrumental in reconstructing the relationship between space and time. Global cities have grown "closer" through increased speed of travel, as well as the intensity of the connectivity among them. The "increasing flows of new kinds" (Ingram, 2005, p. 527) in the space of global networks adds to vulnerability and risk. "Whereas some places are more likely to become conducive to a pandemic disease, there are some travelers who are more exposed as well as more likely to pass the disease on to others" (Aaltola, 2012, p. 63).

Coping with the simultaneity of the crisis: local responses to the pandemic

When the SARS Co-Virus arrived in localities around the world, the major temporal characteristics of this disease became more about simultaneity, including what Rosa (2009) calls "simultaneity of the non-simultaneous." All of the affected countries were forced to respond to the common crisis simultaneously, despite their greatly varied economic, technical, political, and social capacities to do so. The process of time-space compression noted previously significantly reduced, or even annihilated, the time available for the development of the "best" reactions at a local/national level. Even though a synchronized emergency response across countries might be desirable, it is difficult in practice because of the "multi-temporality" associated with the economy, politics, history, and people in different places. We illustrate these issues by focusing on China and Canada, two of the hardest-hit countries, which are geographically distant and became temporally "closer," or interconnected, because of this global pandemic.

After the first SARS case in November 2002, the virus spread unreported in Guangdong Province for a couple of months before moving to Hong Kong and other parts of the world. On March 28, 2003, one day after Beijing was added by the WHO to its list of affected areas, the Chinese Ministry of Health agreed to provide the WHO with regular, up-to-date reports from all provinces beginning April 1. With all other affected areas having already done so, this step symbolized that the country had "become, very clearly, part of the global network in dealing with the disease" (WHO, 2006, p. 24). Although Western media tended to attribute the Chinese government's delay and the inefficient flow of information to its non-democratic system (Huang and Leung, 2005), there were other reasons, too. In addition to China's inexperience in responding to this previously unknown disease, the local health authorities faced barriers, at both systematic (a decentralized system of disease surveillance) and personal (undermining of prospects for job promotion) levels, to the timely reporting of the outbreak to higher authorities. Furthermore, the government's decision on the timing of publicizing the outbreak was influenced by its implications for domestic economic and social stability, and for its already problematic international image (Ahmad et al., 2009; Tai and Sun, 2007).

The huge international pressure to contain the virus induced a high degree of politicization of the battle against SARS in China in order for the government to quickly mobilize the entire society. On April 14, President Hu Jintao declared a "people's war" against SARS (WHO, 2006). On April 20, both the health minister and the mayor of Beijing were removed from their posts (Zhao, 2003). Adopting a strategy analogous to that of the traditional "patriotic health movement" that relies heavily on mass mobilization, China was able to compensate somewhat for the inadequacy of its resources (e.g. time, technology, and personnel) for SARS surveillance and prevention (Liu, 2003; WHO, 2006). Becoming the government's "top priority," SARS responses came directly under the leadership of the vice-premier, Wu Yi, who was *also* made the new health minister. The result was faster coordination of resources, communication and collaboration across sectors (including mass media), places, and hierarchies of the governmental systems (Zhao, 2003; Liu, 2003; WHO, 2006). In late April, it took only seven days for China to build a

1,000-bed hospital for SARS patients in a northern suburb of Beijing, at a cost of 160 million yuan (US$19.33 million) (Zhao, 2003). With the wide dissemination of information and surveillance by the public, the time between onset of symptoms and hospitalization in Beijing was reduced to two days, from the five to six days that was the norm before the outbreak (Pang et al., 2003).

Similar to China, Canada's initial responses to SARS were delayed. As early as November 27, 2002, the Canada-based Global Public Health Intelligence Network (GPHIN), an Internet-based early warning system for worldwide public health threats, received a Chinese-language report of a flu outbreak in mainland China. While the Chinese report was sent to the WHO with a translated English title, the full report was not translated until January 21, 2003, in part because the GPHIN system then in use could not accommodate the information in languages other than English and French (Blench, 2008; Health Canada, 2003). As a result, an early opportunity to learn about SARS was missed by Health Canada and the WHO (Health Canada, 2003). The arrival of SARS in Canada in late February quickly transformed airports into the first and foremost frontiers of disease control. On March 18, quarantine officers were deployed, and "Health Alert Notices" were distributed to air travel passengers arriving in and returning to Canada from Asia at Toronto's Pearson and Vancouver International airports. On April 23, the WHO extended its travel advisory to Toronto, which was removed soon after Canada's strong objection (WHO, 2006). As a condition of this removal, the level of monitoring of passengers at major airports was increased, and multiple expensive thermal scanners were installed at Toronto's Pearson and Vancouver airports in May (Keil and Ali, 2006; Health Canada, 2003). Given the nature of the disease, however, the effectiveness of these airport screening measures was unclear. As of August 27, 2003, out of an estimated 6.5 million passengers screened at Canadian airports, roughly 9,100 were referred for further assessment by screening nurses or quarantine officers: none had SARS. Out of the approximately 2.4 million passengers screened by the pilot thermal scanner project, only 832 required further assessment, and (again) none was found to have SARS (Health Canada, 2003).

On one hand, the massive use of border control regimens, as well as other infection-control instruments (contact tracing and quarantine[1]), for SARS control reflects the decision makers' intention to err on the side of caution and safety in the context of uncertainties (e.g. the difficulty of identifying people with SARS and predicting their movements) (Affonso et al., 2004). On the other hand, however, it also signals "a return to an unhappy past," when "dangerousness" – unpredictable danger that somehow inheres in certain individuals (e.g. people from "Third World" countries) – was the central logic of public health governance, and attention was devoted to "locating and neutralizing all sources of danger" threatening public health (Hooker, 2001, 2006, pp. 179, 180). The transnational movement of SARS was too fast for the classical approach to risk management that required more time to widen "our field of vision" and improve the accuracy of "our prediction" (Cooper,

2006, p. 119). Ironically, at a time when the world has become increasingly deterritorialized due to global capitalism, the importance of border control in disease control was resumed in the context of EIDs.

Being listed by the WHO as one of the "affected areas," and the only such location[2] outside Asia, meant that Toronto also had a hard time adapting to its multiple and contested "temporal" identities. In contrast to a longstanding projected identity as a "safe, North American city," Toronto was constructed during the SARS crisis by the international media as an "exotic plague town," a "backward place" that was "not quite up to modern standards of hygiene and scientific rationality" (Strange, 2006, pp. 221, 222, 223). To protest the WHO's travel advisory, which had tremendous negative impacts on Toronto's image and economy, the governments, along with Canadian media, tried to externalize the disease through, for example, defining the virus as an exotic invasion (SARS as a Chinese or Asian disease), thereby distinguishing Toronto from other affected places (Keil and Ali, 2006; Leung, 2004; Strange, 2006). The city's chief medical officer of health diplomatically commented on Toronto's link with other "pre-modern" SARS-affected Asian areas as "a gross misrepresentation of the facts"; the province's commissioner of public health simply said: "We're not some rinky-dink Third-World country" (Strange, 2006, pp. 224–225).

The racialization of the disease, along with intensified public fear, reactivated historic anti-Asian (in particular, anti-Chinese) and anti-immigrant attitudes in Canada (Keil and Ali, 2006; Strange, 2006). In addition to the surveillance measures targeting travelers from Asia, social distancing from the Chinese and Asian communities in Toronto was widely observed (Leung, 2004). To some extent, the racism during the SARS crisis can also be understood in terms of people's inability to cope with what Harvey (1990) calls "an overwhelming sense of compression," caused by this fast, borderless, and contagious disease. At a material level, the gap between the accelerated global spread of SARS and the lack of effective vaccine and treatment for the disease certainly generated public anxiety and panic. At a cognitive level, reduced travel times between Canada and the "Third World" countries and the permeability of the traditional state borders in the context of EIDs also challenged people's sense of security, given the impaired ability to separate or protect themselves from the "dangerous" others. Although Toronto's ethnic and cultural diversity and transnational ties with other global cities in Asia were negatively presented in the dominant media discourses on SARS, its post-SARS promotional campaigns, ironically, resumed this pre-SARS identity by using the logo, "Toronto: the World Within a City" (Keil and Ali, 2006; Strange, 2006).

Simultaneously confronting the public health emergency, China, as the "source" of the pandemic and a relatively resource-limited country, had to resort to a traditional "patriotic movement" to mobilize its resources; Canada, as a wealthy country "unpreparedly" hit by this disease "from Asia," struggled with the effectiveness of using border control for disease control. We should also note that the mobilization strategy in China was simply impossible in Canada given the latter's decentralized federal system and institutional arrangement of the health care system (Health Canada, 2003; Van Wagner, 2008). These two countries' experiences of coping with the simultaneity of the crisis illustrate the challenges ensuing from the coexistence of and disjuncture between multiple temporalities (Rosa, 2009; Sassen, 2000). These include the accelerated SARS transmission facilitated by the global cities network, when compared with the slower responsiveness of older systems of disease control. Even longer is the time needed for immigrant or diaspora communities to be integrated and accepted into a host country. In the next section, we review the global response to the challenges arising from this newly emergent infectious disease.

Containing SARS at a global level: networked responses in the "spaces of flow"

Our discussion above highlights the rapidity and global extensity of EID infections through air travel built around the global cities network and population density. The same conditions of time-space compression made possible rapid scientific identification of the disease and the development of effective approaches to confining the virus. Once the outbreak began to cross borders in February 2003, it was fully contained within five months. The institutional tools and networks needed for containing the SARS Co-Virus grew out of "speeding up time" in the containment of EIDs and other infectious diseases thanks to several steps taken beginning in the early 1990s. The Program for Monitoring Emerging Diseases (ProMED), launched in 1994, was a first attempt at devising a global Internet-based reporting system on outbreaks of contagious diseases and exposures to toxins. Within 12 years of its founding, it had 30,000 subscribers in 150 countries (Zacher and Keefe, 2008). In 1997, the Canadian government set up the aforementioned GPHIN in cooperation with the WHO. Over time, after overcoming linguistic limitations, it built the capacity to scan news sources 24 hours a day, 7 days a week in Arabic, English, French, Russian, simplified and traditional Chinese, Farsi and Spanish (2008). Meanwhile, six groups of actors took steps to set in place the arrangements needed for rapid construction of response networks for pandemic infectious diseases: the WHO in hiring its own experts; the expansion of trained persons at the six regional offices of the WHO; the involvement and linking together of national research laboratories like the Centers for Disease Control (CDC) in Atlanta and in China; non-governmental organizations (NGOs) like MSF (Médecins Sans Frontières), International Federation of Red Cross and Red Crescent Societies, and Merlin; ministries of health in WHO member states; and other UN bodies like UNICEF (Zacher and Keefe, 2008, pp. 60–64).

Gradually these six types of organization developed an informal network for better communication and responses to problems including disease outbreaks. In April 2000, the network was formalized under the direction of the WHO as the Global Outbreak and Response Network (GOARN). The Network cooperates on four tasks: conducting epidemic intelligence, verifying outbreak rumors, alerting appropriate groups in outbreak situations, and organizing rapid response reactions (Zacher and Keefe, 2008). GOARN has been involved in 70 global outbreaks in 42 countries since its founding: "By assembling missions that rapidly provide critical expertise and resources to countries affected by disease outbreaks, GOARN serves a critical role in containing contagious disease and providing back up capacity" (Ansell et al., 2012, p. 332).

GOARN and SARS: information flows "outpacing" the virus

When compared to the past, acceleration in time "was used to an advantage during the global SARS response" (Ali, 2008, p. 247). Ali adds that the "hallmark of this response was the rapid formation of a virtual network of international scientists who joined forces to identify the causal agent of the disease, develop a universal case definition for the disease, and characterize the genetic code of the virus – all within one month" (Ali, 2008, p. 247).

The first network formed brought together researchers from 13 laboratories in ten countries to tackle the question of the etiology of the virus. One of these laboratories, the British Columbia Cancer Agency in Canada, announced that it had isolated the previously unknown coronavirus and released the data on April 12, 2003. Two days later, the CDC in Atlanta built on the Canadian agency's work, providing further information to the WHO, which, in turn, released the data internationally on April 16. A second network of 50 clinicians in 14 countries developed a definition of the disease and some control guidelines. The WHO set up a third virtual network of 32 epidemiologists from 11 countries, which brought together public health institutions, ministries of health, and WHO country offices to define appropriate public health measures for containing the spread of SARS (Zacher and Keefe, 2008). These networks drew upon the new ICTs to give life to global spaces of flows that linked both nation-states and other transnational actors in the gathering and sharing of information.

Ansell, Sondorp and Stevens (2012) identify additional advantages arising from the acceleration of time and the rapid filling of global spaces by networks of experts. First, it permitted the mobilization of partner institutions as a "technical community," which facilitated rapid coordination due to the "direct and sustained contact" among these institutions (Ansell et al., 2012, p. 324). Second, with such an arrangement, information and influence can flow both down from the WHO but also up from the technical partners to multilateral institutions like the WHO. The independence of the GOARN network added to the effectiveness and efficiency of these transactions. Third, the network form allows the direct integration of quasi-public and even private institutions into virtual pathways for sharing knowledge rapidly and independently. The acceleration of scientific exploration and analysis "via the instant time of computer networks vastly outpaced the biologically defined time of viral reproduction and travel" (Ali, 2008, p. 247). Finally, the accelerated formation of scientific and clinical networks gave an organization like the WHO the capability to manage multilateral, regional and bilateral responses. The WHO's executive director of communicable diseases explained: "I think that it would be fair to say that this is the first global outbreak where there was a 24 hour availability of information and information was continuously coming in through networks of doctors, of clinicians, of virologists, of epidemiologists" (Ali, 2008, p. 242).

Accordingly, working 24 hours a day in "timeless time," GOARN was able to locate and mobilize available and relevant expertise and resources; communicate directly with national ministries of health on needs and terms of reference; mobilize multilateral resources and then deploy them as field teams where needed; and provide a two-way flow of coordination between WHO headquarters and the various field teams involved anytime, anywhere (Ansell et al., 2012).

GOARN was built upon the timeless time that existed between researchers around the world. The WHO was able to challenge the nation-state monopoly on the control of time and thus space. It harnessed the new technologies and their rapid adoption by non-state actors for global public health purposes. As Fidler observes, "[t]he revolution in information technologies changed the context for state calculations about whether to report or try to cover up an outbreak" (2005, p. 346). Ali adds: "As a consequence by 'outpacing' the virus, the scientific establishment was able to break the chain of transmission of the SARS-Co[-]V[irus] quite handily – an outcome that was no doubt assisted by the presence of other fortuitous factors" (2008, p. 247). In this respect GOARN's taking advantage of space-time compression represented a profound break with the state-controlled surveillance found in the traditional approach to infectious disease control in the preceding century. In this regard, GOARN provided additional evidence of the need to reform the International Health Regulations (IHR) built in an earlier world of clock time in limited spaces defined by nation-state and imperial boundaries.

Reforming IHR: toward an accelerated response to future EID outbreaks

Simultaneous to the various steps taken in the 1990s that led to the founding of GOARN in 2000, the WHO had embarked upon the reform of the IHR. These regulations trace their history to the nine sanitation conferences held by various world powers beginning in 1851 and finishing early in the 20th century. Their objective was to minimize the negative impact of several key infectious diseases on international trade, most notably cholera and yellow fever and later influenza. The discussion of how these regulations could be brought up to date was profoundly influenced by the SARS crisis and the emergence of GOARN.

Backed up consistently by the World Health Assembly, the WHO carved out a new approach built on health "security," on the speeding up of time made possible by ICTs, and on the rapid entry of NGOs into global health governance. Over the same period the GOARN model needed to be adapted so as to be able to respond to other threats to global health where the acceleration of time and globalization of spaces were having lethal impacts. The global health discussions now expand to such issues as weapons of mass destruction, biological weapons, chemical disasters like Bhopal in India, and failures of atomic reactors.

Accordingly, when the WHO proclaimed the new IHR regulations in 2005, they took "time and space compression" more fully into account. The particular global health challenges exemplified by the SARS Co-Virus were brought under a new, broader concept – *public health emergencies of international concern*, defined as follows: "an extraordinary event which is determined, as provided in these Regulations: (i) to constitute a public health risk to other States through the international spread of disease; and (ii) to potentially require a coordinated international response" (Fidler, 2005, p. 362). Subsequent legal interpretation of the concept suggests that it would include human-made disasters like Bhopal, the use of weapons of mass destruction, and planned use of biological agents as well as EIDs (Fidler, 2005, pp. 365–367).

The speeding up of time for the spread of EIDs like SARS and the lessons learned about accelerating responses to such outbreaks from the performance of GOARN led to further innovations in the new regulations. They legitimized the practice of states working in global networks in tandem with non-state actors and public health authorities at all levels. The WHO gained the authority to declare these "public health emergencies of international concern" and to issue non-binding recommendations concerning appropriate health measures (Fidler, 2005). "IHR 2005's surveillance strategy … has been specifically designed to make IHR 2005 directly applicable to emerging infectious disease events, which are usually unexpected and often threaten to spread internationally" (Baker and Fidler, 2006, pp. 1059–1060).

Conclusion

ICTs that came together in the 1970s, along with advances in the speed and volume of airplanes, created what Castells (2009) calls "timeless time"; thus many human activities now take place in "no time." Building on Harvey's concept of time-space compression, our chapter argues that contemporary temporality stretches space: the linkages between people are more and more "transplanetary," to use Scholte's term: they reach increasingly from any one place in the world to any other place. These changes exemplify globalization in action and they are most evident along the pathways of the global cities network.

We argue that globalizing processes have changed temporal-spatial dynamics of EIDs like SARS, in such forms as the speed of global transmission, the simultaneity of a public health emergency across geographies, and the possibility of accelerating responses on a global level. Taking into account the intersection of globalization, time and disease, it is clear that contemporary globalization has not only shaped the temporal nature of EIDs like SARS but also has been shaped by the processes of global health responses (e.g. cross-country collaboration and the reform of related international organizations). While the technical infrastructure of global networks has provided a promising condition for accelerating surveillance and information sharing on a global level, other temporality-related challenges – such as differential capacities of the affected countries to respond to the simultaneity of the crisis – are yet to be tackled.

In 2012, the WHO organized an IHR Committee to monitor a new coronavirus, MERS, which first appeared in the Middle East. By 2015, it had expanded to Europe, North Africa, Southeast Asia, China, South Korea, and North America (WHO, 2015). Even more concerning, in 2014, the lethal Ebola virus, which has occurred off and on in rural areas of West Africa for a number of years, reached large cities for the first time, including Lagos, Nigeria, with its 21 million population. Lagos was first classified as a global city (gamma category) in 2008 and had climbed to the "Beta minus" category by 2012. Ebola reaching a global city was sufficiently concerning that the WHO declared the outbreak to be a *public health emergency of international concern* under the IHR in August 2014 (Kennedy, 2014).

In conclusion, the increasing potential for the rapid and global spread of EIDs arising from new temporalities has been our principal concern. The possibility of viruses going *anywhere* in *no time* along global cities pathways heightens tremendously the probability of infectious disease pandemics. The factors that came together in the global outbreak of SARS outlined in this chapter have become even more probable in the ensuing 13 years. More in-depth analysis of the interactions between different temporalities and expanding spaces arising from globalization has become a challenge, if not a necessity, for global health research and globalization studies.

Acknowledgements

This chapter is an updated version of a previously published article in the *Journal of Globalizations*, 13(3), pp. 285–299.

Notes

1 According to Health Canada (2003), about 25,000 residents of the Greater Toronto Area were placed in quarantine during the SARS crisis.

2 Vancouver was removed from the list of "affected areas" three days after the WHO released its first list on March 16, 2003 (WHO, 2006).

References

Aaltola, M. (2012) Contagious Insecurity: War, SARS and Global Air Mobility. *Contemporary Politics*, 18, pp. 53–70.

Affonso, D.D., Andrews, G.J. and Jeffs, L. (2004) The Urban Geography of SARS: Paradoxes and Dilemmas in Toronto's Health Care. *Journal of Advanced Nursing*, 45(6), pp. 568–578.

Ahmad, A., Krumkamp, R. and Reintjes, R. (2009) Controlling SARS: A Review on China's Response Compared with Other SARS-affected Countries. *Tropical Medicine & International Health*, 14, pp. S36–S45.

Ali, S.H. (2008) SARS as an Emergent Complex: Toward a Networked Approach to Urban Infectious Disease. In Ali, S.H. and Keil, R. (eds) *Networked Disease: Emerging Infections in the Global City* (pp. 235–249). Malden, MA: Wiley-Blackwell.

Ali, S.H. and Keil, R. (2006) Global Cities and the Spread of Infectious Disease: The Case of Severe Acute Respiratory Syndrome (SARS) in Toronto, Canada. *Urban Studies*, 43, pp. 491–509.

Ali, S.H. and Keil, R. (2008) Introduction: Networked Disease. In Ali, S.H. and Keil, R. (eds) *Networked Disease: Emerging Infections in the Global City* (pp. 1–7). Malden, MA: Wiley-Blackwell.

Ansell, C., Sondorp, E. and Stevens, R.H. (2012) The Promise and Challenge of Global Network Governance: The Global Outbreak Alert and Response Network. *Global Governance*, 18, pp. 317–337.

Baker, M.G. and Fidler, D. (2006) Global Public Health Surveillance Under New International Health Regulations. *Emerging Infectious Diseases*, 12(7), pp. 1058–1065.

Bashford, A. (2006) "Age of Universal Contagion": Disease, History and Globalization. In Bashford, A. (ed.), *Medicine at the Border: Disease, Globalization and Security, 1850 to the Present* (pp. 1–18). New York: Palgrave Macmillan.

Blench, M. (2008) *Global Public Health Intelligence Network (GPHIN)*. www.amtaweb.org/papers/4.02_Blench2008.pdf.

Budd, L., Bell, M. and Brown, T. (2009) Of Plagues, Planes and Politics: Controlling the Global Spread of Infectious Diseases by Air. *Political Geography*, 28(7), pp. 426–435.

Castells, M.L. (2009) *Communication Power*. Oxford: Oxford University Press.

Cooper, M. (2006) Pre-empting Emergence: The Biological Turn in the War on Terror. *Theory, Culture & Society*, 23, pp. 113–135.

Fidler, D.P. (2005) From International Sanitary Conventions to Global Health Security: The New International Health Regulations. *Chinese Journal of International Law*, 4, pp. 325–392.

Globalization and World Cities Research Network (GaWC) (2012) *The World According to GaWC 2012*. www.lboro.ac.uk/gawc/world2012t.html.

Harvey, D. (1990) *The Condition of Postmodernity: An Enquiry into the Origins of Cultural Change*. Oxford and Cambridge, MA: Blackwell.

Health Canada (2003) *Learning from SARS: Renewal of Public Health in Canada. A Report of the National Advisory Committee on SARS and Public Health*. Ottawa: Health Canada.

Hooker, C. (2001) Sanitary Failure and Risk: Pasteurisation, Immunisation and the Logics of Prevention. In Bashford, A. and Hooker, C. (eds) *Contagion: Historical and Cultural Studies* (pp. 129–152). London and New York: Routledge.

Hooker, C. (2006) Drawing the Lines: Danger and Risk in the Age of SARS. In Bashford, A. (ed.), *Medicine at the Border: Disease, Globalization and Security, 1850 to the Present* (pp. 179–195). New York: Palgrave Macmillan.

Huang, Y. and Leung, C.C.M. (2005) Western-led Press Coverage of Mainland China and Vietnam During the SARS Crisis: Reassessing the Concept of "Media Representation of the Other." *Asian Journal of Communication*, 15, pp. 302–318.

Ingram, A. (2005) The New Geopolitics of Disease: Between Global Health and Global Security. *Geopolitics*, 10, pp. 522–545.

Keil, R. and Ali, S.H. (2006) Multiculturalism, Racism and Infectious Disease in the Global City: The Experience of the 2003 SARS Outbreak in Toronto. *Topia*, 16, pp. 23–49.

Kennedy, M. (2014). WHO Declares Ebola Outbreak an International Public Health Emergency. *The Guardian*, August 8. www.theguardian.com/society/2014/aug/08/who-ebola-outbreak-international-public-health-emergency.

Lee, K. (2003) *Globalization and Health: An Introduction*. New York: Palgrave Macmillan.

Leung, C. (2004) *Yellow Peril Revisited: Impact of SARS on the Chinese and Southeast Asian Canadian Communities*. Toronto: University of Toronto. www.ccnc.ca/sars/SARSReport.pdf.

Liu, C. (2003) The Battle Against SARS: A Chinese Story. *Australian Health Review*, 26(3), pp. 3–13.

Mangili, A. and Gendreau, M.A. (2005) Transmission of Infectious Diseases During Commercial Air Travel. *The Lancet*, 365(9463), pp. 989–996.

May, J. and Thrift, N. (2001) Introduction. In May, J. and Thrift, N. (eds) *Timespace: Geographies of Temporality* (pp. 1–46). London and New York: Routledge.

Pang, X., Zhu, Z., Xu, F.Guo, J., Gong, X., Liu, D., … Feikin, D.R. (2003) Evaluation of Control Measures Implemented in the Severe Acute Respiratory Syndrome Outbreak in Beijing. *JAMA*, 290, pp. 3215–3221.

Rosa, H. (2009) Social Acceleration: Ethical and Political Consequences of a Desynchronized High-speed Society. In Rosa, H. and Scheuerman, W.E. (eds) *High-speed Society: Social Acceleration, Power and Modernity* (pp. 77–112). University Park, PA: Pennsylvania State University Press.

Rosa, H. (2013) *Social Acceleration: A New Theory of Modernity*. New York: Columbia University Press.

Sassen, S. (2000) The Global City: The De-nationalizing of Time and Space. In Ockman, J. (ed.), *The Pragmatist Imagination: Thinking About "Things in the Making"* (pp. 254–265). Princeton, NJ: Princeton Architectural Press.

Sassen, S. (2001) *The Global City: New York, London, Tokyo*. Princeton, NJ: Princeton University Press.

Scholte, J. (2005) *Globalization: A Critical Introduction* (2nd edn). Basingstoke and New York: Palgrave Macmillan.

Strange, C. (2006) Postcard from Plaguetown: SARS and the Exoticisation of Toronto. In Bashford, A. (ed.), *Medicine at the Border: Disease, Globalization and Security, 1850 to the Present* (pp. 219–239). New York: Palgrave Macmillan.

Tai, Z. and Sun, T. (2007) Media Dependencies in a Changing Media Environment: The Case of the 2003 SARS Epidemic in China. *New Media & Society*, 9, pp. 987–1009.

Taylor, P.J. (2004) *World City Network: A Global Urban Analysis*. London: Routledge.

Taylor, P.J. (2013) *Extraordinary Cities: Millennia of Moral Syndromes, World-systems and City/State Relations*. Cheltenham: Edward Elgar Publishing.

Van Wagner, E. (2008) Toward a Dialectical Understanding of Networked Disease in the Global City: Vulnerability, Connectivity, Topologies. In Ali, S.H. and Keil, R. (eds) *Networked Disease: Emerging Infections in the Global City* (pp. 13–26). Malden, MA: Wiley-Blackwell.

Weiss, R.A. and McLean, A.R. (2005) What Have We Learnt from SARS? In McLean, A.R., May, R.M.Pattison, J. and Weiss, R.A. (eds) *SARS: A Case Study in Emerging Infections* (pp. 112–116). New York and Oxford: Oxford University Press.

World Health Organization (WHO). (2006) *SARS: How a Global Epidemic was Stopped.* Manila, The Philippines: WHO Regional Office for the Western Pacific Region.

World Health Organization (WHO) (2015). Coronavirus Infections, June 12. www.who.int/csr/disease/coronavirus_infections/en/index.html.

Zacher, M.W. and Keefe, T.J. (2008) *The Politics of Global Health Governance: United by Contagion*. London: Palgrave Macmillan.

Zhao, Q.-Z. (2003). A Hard Won Victory by China and its People. *People's Daily*, July 1. english.peopledaily.com.cn/200307/01/eng20030701_119229.shtm.

8 The inertia of energy

Pipelines and temporal politics[1]

Brent Ryan Bellamy

> Every round of new pipelines and tankers and deep-water drilling rigs encumbers the next decades with an even more ponderous mass of infrastructure into which carbon has been locked: the ruts of path dependency deepen. Every generation presiding over growing emissions adds more than the former to the accumulation of CO_2 in the atmosphere. For every year global warming continues and temperatures soar higher, living conditions on earth will be determined more intensely by the emissions of yore, so that the grip of yesteryear on today intensifies – or, put differently, *the casual power of the past inexorably rises*, all the way up to the point when it is indeed "too late." The significance of that terrible destiny, so often warned of in climate change discourse, is the final *falling of history into the present.*
>
> (Andreas Malm, *Fossil Capital*, p. 9, original emphasis)

The inherent question of this volume, the relation between time and globalization, finds form in the oil pipeline. Snaking across Africa, Asia, Europe, the Middle East, and North and South America, miles of pipe take shape in order to convey fossil fuels from the source of extraction to refineries and distribution centers. Pipelines are part of a mechanism in fossil capital that generates an experience of the pure present. So long as oil keeps flowing, the length and duration of the journey does not matter – each instant when oil is fed into the mouth of the pipeline is completed in the barrel at the other end.[2] Without an adequate replacement or an alternative way of organizing the social, fossil fuels continue to be the driver of contemporary life. Pipelines are the mechanism that keeps the motor running. But what kind of future does pipeline construction build?

As the epigraph bears out, fossil-fuelled global warming is an historical event unlike any other in that it is intensely recursive. Thus, I consider oil pipelines as instruments that have profound impacts on the experience of the present and on how others might experience the future. While it may address supply side problems, this infrastructure brings future costs with it that do not appear in its schematics. In this chapter I explore two modes of resistance to pipeline development in order to better frame the ideological problems posed by high-density energy. Here, the pipeline becomes a tool for understanding and critiquing fossil capital, while modes of oil-infrastructure resistance offer further insight into the scope and velocity of the inertia of energy.

Pipelines came into being as part of what Fredrick Buell calls the "bootstrapped system" (Buell, 2012, p. 280) of fossil capital transport: first carts and barrels, then rail, and then pipelines.[3] Unlike those earlier modes of conveyance, the pipeline appears shaped by the logic of fossil capital. That is, the pipeline is shaped in the interest of a smoothness that does not require workers and can flow evenly throughout the entire day. In *Carbon Democracy*, Timothy Mitchell reveals a key political distinction between the movement of coal and the flow of oil: coal follows "dendritic networks with branches at each end but a single main channel," which creates "choke points at main junctures"; oil flows through networks that have "the properties of a grid" with multiple possible paths, so that "the flow of energy can switch to avoid blockages or overcome breakdowns" (Mitchell, 2011, p. 38). Crucially, Mitchell continues, "these changes in the way forms of fossil energy were extracted, transported and used made energy networks less vulnerable to the political claims of those whose labour kept them running" (Mitchell, 2011, pp. 38–39). The politics of the age of coal, such as the strike and the blockade, do not obtain in the oil age. Other forms of resistance must be necessary to create the friction necessary for transition to other forms of energy or new, less intensive energy habits.

As an infrastructural innovation of fossil capital, the pipeline mediates the complex relation between oil's spatial infrastructure and its movement from the point of extraction to the point of refinement to the point of consumption. It enables trajectories to be cut across space and through time. Oil's mobility as substance and the way it enables movement fundamentally informs social and economic life.[4] The vast reliance, in modernity and the contemporary moment, on energy abundance is written across the landscape in the steel used to create literal flows of long-dead, high-energy matter across contested landscapes, settled terrain, and the space of capital.

The pipeline also makes bids on futurity. The continued use of oil implies that the future will continue to be the same as the present. Oil fuels manufacturing automation and our vehicles and heats our homes and offices; fossil-fuel derivates, like plastics, are still the basis of our commoditized lives. As Imre Szeman puts it:

> From oil flows capitalism as we still know it: the birth of the first giant multinationals – Standard Oil (whose component elements still persist in Exxon Mobil, Texaco, and British Petroleum), DuPont, and the Big Three automobile makers; the defining social system of private transportation – cars, air travel, freeways, and with these, suburbs, "white flight," malls, inner-city ghettoization, and so on; and the environmental and labor costs that come with access to a huge range of relatively inexpensive consumer goods, most of which contain some product of the petrochemical industry (plastics, artificial fibers, paints, etc.) and depend on the possibility of mass container shipping.
>
> (Szeman, 2007, p. 806)

Szeman's larger claim is that from oil capital flows different approaches to the imaginable future: the current conjuncture has been imagined through politics and "strategic realism," science and "techno-utopianism," or catastrophism and "eco-apocalypse" (Szeman, 2007, p. 808). I would add to Szeman's assessment that the way we imagine oil futures gets embedded in oil infrastructure. The oil pipeline enforces a temporal coevality, affirming that all future people or cultures are contemporaneous with the present. However, for those watching the diminution of Arctic and Antarctic ice sheets, the assertion that the future is contemporary with the present registers as a fiction that masks (or simply ignores) our current climate crises. We still drill, frack, pump, refine, ship, and consume fossil fuels; how to resist these economies, processes, and their imaginaries becomes a central question today.

Further, one might ask, what happens to pipelines when they no longer prove useful? Cultural theorist Stephanie LeMenager describes the act of abandonment, "partial removal, or toppling," of oil and gas pipelines in the vernacular as "reefing" (LeMenager, 2014, p. 180). Whether in use or not, pipelines, and their skeletal remainders, persist. LeMenager suggests that the oil industry and federal government save money from this ease of abandonment and, further, that the question of "what to do with unusable pipelines is rarely broached," as "most of these pipes inadvertently get reefed" (LeMenager, 2014, p. 180).[5] LeMenager's assessment has two implications for this chapter. On the one hand, she draws out the difficulty of imagining energy infrastructures. The example of the reefed pipelines shows that it remains difficult to think beyond the local or the regional to imagine pipelines that cover North America or Central Europe. Indeed, despite best efforts there is no global map of our energy infrastructures.[6] On the other hand, she casts light on the operations of the political economy of energy transportation as a site of struggle for the future. While legislation makes it easier for oil companies to operate, entrenched legal structures simultaneously make it more difficult to work toward a future beyond or after oil.

This essay explores the consequences of our current energy commitments through the formal structure of the pipeline; my work falls under the guise of what scholars have been calling the cultural politics of oil, or the study of petrocultures.[7] Thinking about petroculture, simply put, involves giving energy, specifically petrol, a central role within humanities and social science frameworks. The benefits of this approach outweigh the risks. While cautioning against reading energy as the prime mover of history, Allan Stoekl writes: "the most effective way of refusing such a reification of oil, all the while granting it the visibility it deserves, is to write its history … It's when we think about what 'oil history' could mean that we take a *natural* entity and recognize its *cultural* centrality" (Stoekl, 2014, p. xii). Though oil presents itself as critically overwhelming, responses to it should find ways to *mediate* the particulars of oil and the general situation of our energy system – a task which increases in difficulty and immensity when considering its inherently global nature. In this light, petroculture becomes inseparable from the history of modernity and globalization as it helps to accomplish what Marx, in the *Grundrisse*, and (as Rebecca Solnit points out) so many others have called the annihilation of space by time.[8] However, as this collection bears out, time has so frequently been omitted from conversations about globalization and late capital, which are often reduced to a spatial account of the pure present.

Thus, my aim in this essay is to take the pipeline as an infrastructural innovation of petroculture and examine both its implied temporality – a future that resembles a kind of endless, oil-infused present – and alternate temporalities that resist or stave off the future of oil, either in the name of some other energy culture or in the wake of the petro present's inevitable collapse. As we sink further into the sedimenting mire of our own inopportune need for energy, we become stuck with the lives and afterlives of our energy infrastructures. So, in what follows, I turn to the juridical slowing of legal inquiry and temporal negotiation of literary speculation as lifelines, as alternate temporalities that could potentially interrupt the rhythm and tempo of inertia. These alternatives involve or deploy a slowing of time coupled with an assertion of difference, whether articulated as a relation to land or in the dialectic of historical science fiction. Following the work of Mark Simpson on oil's fantasy of "lubricity," I consider the slowing of petroculture in terms of its *political friction*. In Simpson's language, interventions in the smooth flow of oil reckon "friction's differential power."[9] While holding development in abeyance these slowed temporal moments also offer purchase for those wishing to act in the interest of an alternate energy future.

This is not a history or political economy of the pipeline; instead, I frame it as a cultural form of fossil capital in order to posit a set of other forms that oppose and potentially work against the dominance of the pipeline. These two moments of resistance work to dramatize the dynamic Andreas Malm addresses in the epigraph. In the case of fossil capital and global warming, the falling of history into the present is at once global and temporal. The legal struggle against the Mackenzie Valley Pipeline project in northern Canada offers a temporal alternative to the inertia of energy. This episode is illustrative of wider global problems relating to the political economy and environmental politics of resource extraction and Indigenous rights. Meanwhile, the future history that Naomi Oreskes and Eric M. Conway offer intervenes in the imaginary of energy infrastructure and offers a chance to reflect on what kind of future readers might be capable of imagining. I now turn to the contested territory of the Canadian north in the mid-1970s.

The Mackenzie Valley Pipeline and the contest over Indigenous land

On March 21, 1974, the government of Canada commissioned Justice Thomas Berger to investigate the impact of a pipeline proposed by Canadian Arctic Gas Pipeline Limited that would run up the Mackenzie Valley crossing into the northern Yukon from the North West Territories. At the time, pundits heralded the project as the largest private construction enterprise in history. In *The Past and Future Land* (1976), journalist Martin O'Malley emphasizes that never before had so large an undertaking as the Mackenzie Valley Pipeline been proposed, nor had an inquiry into the development of such a project been undertaken (the inquiry ended up costing more than C$5 million) (O'Malley, 1976, p. 11).[10] What became known as the Berger Inquiry appears, in the present, as a wonderful yet singular event in the history of clashes between Canadian oil and gas development and Indigenous rights and land claims.

In early March 1975 three lawyers from Toronto spoke before the inquiry: one on behalf of the inquiry, one on behalf of the pipeline company, and one on behalf of the Indigenous, Inuit, and Métis peoples. Each stood for his respective party's interests: for openness, for economic activity and progress, or for the interests of the peoples of the Mackenzie Valley. Métis lawyer Glen Bell argued that in "the process of building a pipeline, native society will be dragged from a land-based economy – hunting, trapping, and fishing – into a wage economy." He further claimed that the "big petroleum companies will move in and suck dry the reserves of gas and oil from the north," asking, "after they have gone what happens to the native people who must continue to live there? What will be left for the owners of the land after the Canadian Arctic Gas and its pipeline are just a memory?" (O'Malley, 1976, p. 3). Glen Bell's commentary suggests a time other than that of development and construction, and frames, in settler language, the stakes of our energy commitments. Talking and listening, in the case of the inquiry, model one form of resistance to the powerfully dominant time of oil capital and a way to work against the bid that an energy-intensive present places on the shape of the future. The rhythm and duration of a long conversation proved effective precisely for interruption and replacement of a settler imaginary relationship to the north.

The inquiry took place in the midst of the 1970s energy crisis when both developers and government were investigating ways to develop and deliver new supplies of fossil fuels. According to legal historian Kim Stanton, "significant oil and gas deposits had been identified in the western Arctic in the late 1960s," resulting in industry players forming and submitting proposals for a pipeline that would run through the western Arctic and western provinces to the border and into the United States (Stanton, 2012, p. 84). These proposals met with simultaneous support and opposition. In 1972, the federal government tabled a document, *Expanded Guidelines for Northern Pipelines*, which supported the pipeline; at the same time, Indigenous rights organizations and environmental activists expressed their opposition to the project. Martin O'Malley carefully spells out the future embedded in the approval of the pipeline: "If it is approved, the pipeline would be 'looped' within five years, which really means a second pipeline. After that, gas, oil, and minerals explorations will be stepped up, then an oil pipeline, hydroelectric transmission lines, a highway, telecommunications, and possibly a railway" (O'Malley, 1976, p. 13). O'Malley's description of northern development fits with Buell's language of the "bootstrap" effect, where one infrastructural project always entails an extended network of other required commitments – a settler version of the future north.

In the language of geographer David Harvey, this bootstrap effect could also be called a *spatial fix*: on the one hand, a chance to allow investors to create jobs in the north, thus fixing or mending the economy and resolving a potential crisis; and on the other, a chance to set in place, *to fix*, the infrastructure of fossil capital (Harvey, 2003, pp. 87–89). In temporal terms, the building of a pipeline is a way to maintain the dominance of the order of the present: in the terms of both settler colonialism and fossil capital. From this vantage point, the spatial fix also appears to be a temporal one: the pipeline reproduces dependence on a particular formation of oil capital and, in a settler-colonial context, violently severs Indigenous connections to the land. The Mackenzie Valley Pipeline proposal thus shuttles in more than simply one pipeline; instead, it suggests a full-scale development of the traditional lands of the Dene, Inuit, and Métis peoples.

The thrust of development in the north is by no means a recent phenomenon. O'Malley also looks to the past and contends that pipelines are merely the most recent attempt to settle the north. He adds energy surveyors to a long list of other developers and prospectors. The list includes "explorers, whalers, fur traders, white trappers, missionaries, highway-builders, DEW-line workers, gold-diggers, assorted anthropologists, geologists, and archaeologists, as well as a few hippies and intrepid tourists." O'Malley continues, "[b]oomers come and go but natives stay in the north and die in the north and when the non-renewable resources that are in such demand are sucked dry or blown out of the ground, the natives will still be there" (O'Malley, 1976, p. 13). This static view of indigeneity reinforces O'Malley's claims, even as it leaves aside the economic, environmental, and social consequences that this level of development would bring along with it. The development of a large-scale energy infrastructural project commits whole regions to a petrocultural way of being and knowing. The development of the commission provided a platform to those opposed to the project, especially the Dene, Inuit, and Métis people of the Northwest Territories. The recognition of Indigenous voices on the issue remains of the utmost importance, yet the political impact of the slowing that the 21-month commission created should not be overlooked.

Through its attention to Indigenous peoples and lands, the Berger Inquiry generates political friction that is anathema to the energy companies' inertial pace of development. As a juridical enterprise of consultation, the Berger Inquiry created an opening for Indigenous ways of knowing and deciding to operate. It made time for people to express their own desired future. The people of the Mackenzie Valley came out in great number to assert their claims, which forestalled land development. By including Indigenous ways of governance, the commission effectively slowed down the progress on construction. It created a temporality based in the realm of speaking and listening. Communication at these speeds was paramount for a proper and complete fulfillment of its charge – to hear what the residents of the north thought about the pipeline and how it might affect their lives. This conversation created a pause for the development of the north and allowed those involved to test various future possibilities, to encounter, in the realm of the imagination, different futures. These futures cannot be thought of in terms of reaction, conservation, development, and progress, but instead need to be conceived as alternate temporalities from the very social form that shapes those political options in the first place. Consulting the Dene and the Inuit, or other Indigenous people in the north, meant considering a different way of being and a different way of knowing the land. Different from one another, certainly, but also different from the relationship between humans and the land that capital produces.[11] Indigenous relations to the land, as Dene scholar Glen Coulthard has so recently been reminding us,[12] are central to Indigenous ways of life and also to resisting the dominance of settler-colonial capitalism.[13]

The pipeline – whether aborted as in the case of the Mackenzie Valley project, or actively contested in recent developments such as the Northern Gateway Pipeline – is a violent tool for the dispossession of Indigenous lands.[14] It slices the landscape, cutting a trajectory and drawing a boundary. These boundaries do not just obtain in space, but also apply in the time of national belonging, with its electoral cycles and bureaucratic methods, and in the time of capital, with its redefinition of the land as a free gift and human labor as a commodity. In this way, globalization and the petro-present leave incredible, indelible marks on attempts to revitalize the Indigenous movements of the past and present or attempts to imagine a future after oil. Activating the slow time of juridical debate and consultation remains difficult, and by no means changes the current order of things.[15] This is not to say that the struggle is futile, but rather to suggest that an intervention also needs to be made at the level of the cultural imaginary.

The collapse of Western civilization and the oil imaginary

Petroculture demands that we conceptualize time differently. It recalibrates our conception of past, present, and future, through what Fredrick Buell has called the narratives of exuberance and catastrophe. For Buell, these two narratives are mutually dependent components of fossil capitalism: in the good boom years of oil's dominance everything seems plentiful, and the contemporary sense of catastrophe is deeply entwined with this exuberance. From the soaring highs of abundance to the crashing lows of shortage, each moment in this two-step dance has a mode of temporality. Energy exuberance tends to enable imaginaries of the future and re-visions of the past like we see in 1960s cartoons such as *The Jetsons* and *The Flintstones* – things are looking up and they have basically always been good. This temporal flattening creates a cultural and economic situation that is ripe for the disruption of a crisis that strikes hard and fast. Whether things are looking up or teetering on the edge of crisis, each phase brings with it a set of boundaries for imagining the future. This is not to say that one can only imagine plentiful bounty in good times and massive destruction in bad, but rather that these appear to be the most accessible and plausible forms of future thinking available at each given time.

The Collapse of Western Civilization (2014) by Naomi Oreskes and Erik M. Conway, two historians of science, surely marks our current moment as one of catastrophe. The book, which I hesitate to call a novel, combines the science-fictional conceit of an imagined future with a rigorously thought historical document in order to come to terms with the contemporary energy-ecology impasse.[16] The look backwards from the imaginary future of 2393 offers a way to come to terms with the cognitive inertia of energy infrastructure through the estrangement effect of a distant future. The book uses the *longue durée* as a structuring conceit, extrapolating the effects of a global time lapse to identify the deep-seated reasons behind our present destructive commitments. This negotiation of time, especially the way it is presented as rapidly passing before the reader's eyes, has the effect of returning the reader to the present. In its three short chapters, *Collapse* pushes the reader to imagine just what energy commitments in the present mean for the future, and precisely why it is so urgent not only to rethink those commitments, but also to rethink the whole network of scientific practice and economic doctrine that shore up oil capital. The temporal maneuvers of the narrative offer more incisive and provocative remarks about the present than would seem possible for scientific articles. Science fiction's capacity to develop a critical understanding of the present, here, plays a key role in this alternative to the inertia of energy.

Oreskes and Conway use the device of the scientist-narrator to address an imagined future audience while simultaneously targeting ways of imagining the present. For instance, they distance the imagined future through the plain-spoken asides of the narrator. The narrator explains the politico-geographic language he uses – terms like "nation-state" and "Cambria" or "the Nordo-Scandinavian Union" – to teach the reader how to understand the extrapolation at work in the text (Oreskes and Conway, 2014, p. 15). In effect, the narrator explains the world of Oreskes and Conway's present, the reader's present, from a scientific-political angle that renders visible the systems that we so quickly take for granted in the same way that they are visible from the standpoint of oil – as mere barriers to the smooth transportation of high-density crude.

Collapse offers a history of science leading up to the present of the narrator, which, at the same time, is partially the reader's history and partially the reader's future. The scientist tracks the development of anthropogenic climate change and people's awareness of it, from the discovery of CO_2 to the rise of environmental movements, and from the foundation of the Environmental Protection Agency to that of the Intergovernmental Panel on Climate Change (Oreskes and Conway, 2014, pp. 16–18). Alongside the hard facts of scientific discovery, the narrator tracks the problem of climate-change denial. This history hinges on a kind of cognitive inertia, though the book itself gives it a different name. The narrator traces climate denial to the United States, noting that "some countries tried but failed to force the United States into international cooperation. Other nations used inertia in the United States to excuse their own patterns of destructive development" (Oreskes and Conway, 2014, p. 18). The text itself leaps over this impasse in order to continue to map the overdetermined sources of energy's inertia.

The narrator then describes events as they unfold in the 21st century, mentioning a series of artworks that endured through the heated debates between scientific truths and the cash spent to buy scientific opinions.[17] For instance, in 2025 the United States introduced a National Stability Protection Act (modeled on the Sea Level Rise Denial Bill) which led to "the imprisonment of more than three hundred scientists for 'endangering the safety of and well-being of the general public with unduly alarming threats'" (Oreskes and Conway, 2014, p. 25). The narrator explains that this kind of response is described in his time as a part of the psychological phenomenon known as "*human adaptive optimism*," which could be read as the mirror image of *cognitive dissonance* – the making of choices against one's own understood best interests. Compounding with the brutal denial of scientific fact, these limits to human action also imply that the right cultural form might begin to instill the kind of critical understanding necessary to move beyond simply knowing the problem. The text traces these explanations to their root in Enlightenment thought. They call this impasse the fallacy of Baconianism, where people do not take corrective action despite their knowledge of threats. Unlike the Roman and Mayan empires, where historians, archaeologists, and synthetic-failure paleoanalysts are unable to agree on the engine of their ruin, in the case of the 21st-century nation-states that referred to themselves as *Western civilization*, experts agree that people "knew what was happening to them but were unable to stop it" (Oreskes and Conway, 2014, pp. 14–15). The narrator refers to "the archaic Western convention of studying the physical world in isolation from social systems," as evidenced in the very title of the position, "physical scientists." In this manner, they develop an Enlightenment legacy – the impasse of knowledge and action – while insisting on the need to bring physical scientists into conversation with the social and humanistic sciences.

The second chapter of *Collapse*, "The Frenzy of Fossil Fuels," grapples directly with the cognitive inertia of energy commitments. The narrator asks: "How did these wealthy nations – rich in the resources that would have enabled an orderly transition to a zero-net-carbon infrastructure – justify the deadly expansion of fossil fuel production?" (Oreskes and Conway, 2014, p. 35). The initial answer to this question has already been explained above: denial. The second answer – that shale gas deposits could offer a bridge to renewable energy – shows a careful negotiation of the present political economy. The narrator addresses this answer by outlining its fallacies: fugitive emissions, a poor distribution system, replacing the wrong fuel sources, a sustained low price for fossil fuels prevents new energy sources from emerging, and so on (Oreskes and Conway, 2014, pp. 36–37). The text highlights the result of not breaking free from this inertia – crippling heatwaves, crop failure, rising ocean levels, and changing weather patterns. Here the temporal device works in the favor of the text as it avoids turning into an apocalyptic thriller. The narrator claims, "there is no need to rehearse the details of the human tragedy that occurred; every schoolchild knows of the terrible suffering" (Oreskes and Conway, 2014, p. 44). In a suspenseful bid, Oreskes and Conway show the reader more by leaving it up to his or her imagination. Fossil fuels, the path of least resistance, are also the path that could prove humanity's ultimate failure to adapt to a world of their own making.

The work of defamiliarization is further developed through maps included throughout the text. Each chapter starts with a regional or a continental map that depicts ocean levels in one geographic space across two temporal planes. The reader will recognize a map of his or her own present, and will not recognize, superimposed in a ghostly imprint, the land that *was* – that *will be* – swallowed by the rising oceans.[18] At the start of the text, the map of the former Netherlands places Brussels on the coast, while it submerges The Hague, Antwerp and Amsterdam, and the highlands around Apeldoorn become an island. The second map swallows the floodplains of Bangladesh, and the third depicts what is left of Manhattan and the boroughs of New York City as a new string of thin islands. This device works to reinforce the temporal negotiation of the text itself, hammering home the ecological imperative to imagine and move toward a world beyond oil. Readers can look at the land that will be under water and consider their own relationship to those places, not only geographically, but also temporally. In this way time becomes the central axis on which the political bid of *Collapse* turns, as the reader is drawn into imagining the future only to reflect again on the present. The loop of time is what effects the kind of slowing that I earlier identified with the grounded norms of Indigenous cultures, only here the loop is created through the estranging effects of science fiction.

Entropy as politics

In order to abandon the commitment to fossil fuels, we must turn away from the most energy-intensive development in human history. As James Marriot and Mika Minio-Paluello persuasively argue in *The Oil Road*, "[t]he solution to the unsustainable extraction of oil and gas – from both an economic and an environmental perspective – is simple: stop drilling for oil and gas" (Marriot and Minio-Paluello, 2013, p. 47). I would add to this, the solution to breaking free of the inertia that comes along with energy infrastructure is simple: stop building rigs, pipelines, and refineries, and start re-imagining what those structures can be and do. Exploring our energy histories and extrapolations, then, offers us a chance to investigate our present troubles. Energy inertia might as well be the impasse that our energy infrastructures get us into, but what can be done now to creatively redeploy those infrastructures toward new ends? The temporal politics of slowing, as explored in the Berger Inquiry and *The Collapse of Western Civilization*, allow this kind of question to be asked.

However, these texts do more than slow down the time of oil development; they also zoom out into a more expansive sense of time, within which the pipeline can be seen as a short-term event with devastating consequences (the rhetoric surrounding oil infrastructure and job creation appears particularly naive in this light). On the one hand, the Berger Inquiry takes seriously Indigenous sovereignty and the grounded normativity of Indigenous peoples who have walked these lands since time immemorial, while on the other, *Collapse* speeds up the passage of time, effectively transporting the reader several centuries into the future almost instantaneously. The book zooms forward through the time of oil, offering a suddenly expansive vantage point from which to observe our current dependence on oil as a dangerous quirk of history, rather than as a permanent, endless inevitability. These adjustments to the temporal scale of globalization, to the time of oil capital, offer a crucial moment to consider alternative relationships to time and to energy.

In "Logistics and Opposition," radical philosopher Alberto Toscano implores progressive thinkers to engage with the question of infrastructure.[19] He undoes the kind of work that urges rupture and blockage above all other avenues for political action. Toscano writes: "The question of what use can be drawn from the dead labours which crowd the earth's crust in a world no longer dominated by value proves to be a much more radical question, and a much more determinate negation than that of how to render the metropolis, and thus in the end ourselves, useless" (Toscano, 2011). He turns to the electrical grid as a site where "mere sabotage" falls short of the potential for anti-capitalist politics, asking radicals to consider its refunctioning in a way that it is "*pre-communist* rather than post-modern" (Toscano, 2011). Building on Toscano's practical and seductive suggestion, I would urge us to consider alternate futures for our energy infrastructures.

Let me conclude, then, with a call to imagine alternatives. While the Berger Inquiry remains an inspiring model for a conversation between Indigenous peoples and settlers that should never have stopped, the conversation could be taken further than preventing development in the north and toward decolonization – for instance, Coulthard offers a compelling account in his book about the prefigurative politics of direct action.[20] Furthermore, we see an inspiring development in the Indigenous mass movement Idle No More, whose very name addresses the relationship between inertia and politics. Founded in December 2012, its vision is to call "on all people to join in a peaceful revolution, to honour Indigenous sovereignty, and to protect land and water" (Idle No More, n.d.). The impetus of the name and the movement built up in response to bill C-45 – a Canadian Conservative government omnibus bill that threatened Indigenous rights. What started with Nina Wilson, Sheelah McLean, Sylvia McAdam and Jessica Gordon organizing a teach-in continues to grow today.[21]

Meanwhile, Naomi Oreskes and Eric Conway provide a place to start imagining how the next 50 years could play out if we do not break free of energy inertia. Together, resistance to the inertia of energy must be taken up both in terms of ongoing settler colonialism and the cognitive blockages of our petroculture. It remains up to us to find new ways to imagine the impasse and the way beyond it. Pipelines may impose incredible temporal limits on imagining a different globalization, but in spite of these limits, we need to seek solutions that use the form of our energy infrastructures to imagine a new global world, and other, less intensive, energy futures. Cultural theorist Jeff Diamanti recently offered a compelling formulation in "Three Theses on Energy and Capital," in which he suggests that the difference between energy and capital is that "*while capital in circulation picks up steam, energy loses it*" (Diamanti, 2016, p. 14, original emphasis). What we need today is not inertia, but the temporal slowing of friction: in order to create the time necessary to imagine an alternate energy future, it would seem we need to embrace entropy itself.

Notes

1 I want to acknowledge the Infrastructure Working Group, which formed at the 2013 Banff Research in Culture residency, for getting me thinking about infrastructure as a cultural form. Thank to you Shama Rangwala for her insightful comments and astute editorial suggestions. Thank you to Mark Simpson and Darin Barney for sharing their forthcoming work with me. I am indebted to Imre Szeman and Sheena Wilson whose work on energy and culture remains inspirational.

2 Whether by rail, ocean tanker or pipeline, the energy costs of transporting oil are great.

3 "Before the development of extraction techniques came experiments with refining oil and the development of lamps suited for its use as a luminant. Also before extraction, capital accumulation began and marketing was pioneered, two other crucial parts of the oil converter chain. And together with extraction, storage and transportation converters had to be immediately developed – and go through many phases, as teamsters hauling carts with barrels yielded to railroad tankers and then to pipelines" (Buell, 2012, p. 280).

4 See Huber, 2013; Johnson, 2014; and LeMenager, 2014.

5 One environmental problem with reefed pipelines is that they do not tell their stories, so that anyone happening upon one would not be able to tell whether it still had residues of oil or gas by-products lurking stagnant within it. Meanwhile, her focus remains local: LeMenager is only talking about the 31,000 miles of pipelines that are buried beneath the sea floor in the Gulf of Mexico.

6 Ann Chen, a 2014/15 Fulbright Visiting Research Fellow in the English and Film Studies Department at the University of Alberta, will be recording and mapping the "cultural, political and environmental zeitgeist, on the cusp of change," as she asks "who or what stands to be changed by the construction and operation of this pipeline?" (Chen, 2014a). Chen's project blog, "Pipeline Mapping," can be followed here: pipelinemapping.tumblr.com. Cf. Insko, n.d.; and Ferguson, n.d.

7 "*Petrocultures* is a new research cluster at the University of Alberta whose aim is to support, produce, and distribute research related to the social and cultural implications of oil and energy on individuals, communities, and societies around the world today. University of Alberta researchers are situated in a prime location to observe, assess and analyze the multiple and complex impacts of the development and management of the oil industry and of energy more generally. The research activities and structures created by the Petrocultures Research Cluster will

enhance and expand this research, and position the U of A at the forefront of a growing field of academic study" (Petrocultures, 2013).

8 "Capital by its nature drives beyond every spatial barrier. Thus the creation of the physical conditions of exchange – of the means of communication and transport – the annihilation of space by time – becomes an extraordinary necessity for it. Only in so far as the direct product can be realized in distant markets in mass quantities in proportion to reductions in the transport costs, and only in so far as at the same time the means of communication and transport themselves can yield spheres of realization for labour, driven by capital; only in so far as commercial traffic takes place in massive volume – in which more than necessary labour is replaced – only to that extent is the production of cheap means of communication and transport a condition for production based on capital, and promoted by it *for that reason*" (Marx, 1973, pp. 524–525). Cf. Solnit, 2003.

9 See Simpson, n.d., forthcoming.

10 O'Malley suggests that "the Berger Inquiry is the product of unnatural circumstances, among them a minority government in 1972 and a Pierre Trudeau on the defensive," and reveals that a "source close to the inquiry doubt[ed] that the government realized what it was unleashing when it appointed Berger to head the inquiry" (O'Malley, 1976, p. 12).

11 See Foster, 1999.

12 As a settler, I include myself in the "us," here, even as I recognize that those reading this could certainly need no reminding.

13 In *Red Skin, White Masks* (2014), Glen Coulthard locates a site of Indigenous resistance to settler colonialism in what he calls *grounded normativity*: "When related back to the two pillars of Marx's primitive accumulation thesis – dispossession and proletarianization – it would appear that in Canada the history and experience of the former has structured the political relationship between Indigenous peoples and the state to greater extent than the latter. I also suggested that the primary experience of dispossession is what tends to fuel the most common modes of Indigenous resistance to and criticism of the colonial relationship itself: that is, Indigenous struggles against capitalist imperialism are best understood as struggles oriented around the question of land – struggles not only *for* land, but also deeply *informed* by what the land as a mode of *reciprocal* relationship (which is itself informed by place based practices and associated forms of knowledge) ought to teach us about living our lives in relation to one another and our surroundings in a respectful, nondominating, and nonexpoitative way" (Coulthard, 2014, p. 60).

14 See Greene, 2014; and Hui, 2014.

15 Socio-cultural anthropologist Pamela Stern identifies a cultural shift from the time of the 1970s, when Berger's Inquiry attracted the attention of the nation, to the present Mackenzie Valley Pipeline Proposal that has provoked little public discussion. See Stern, 2007. Cf. cultural critic Sheena Wilson's assessment, in "Gendering Oil," of both the prominence of the Idle No More movement and the way gender seeps into commentary on environmental messages and activisms. One of her examples is the way Chief Theresa Spence, an advocate of Idle No More, became discussed and represented during her hunger strike in December 2012. Wilson argues that "the violated female body, in association with the Idle No More movement, functions to spectacularize and market petro-violence to the media-consuming public as a reaction to the Idle No More resistance movement rather than as endemic to the legacies of colonial logics, human rights abuses, and gender-sexual violence on which Western petrostates are founded" (Wilson, 2014, p. 245).

16 The text takes the occasion of the fictional "tercentury of the end of Western Culture (1540–2093)" to address the incredible failure to act on "robust information about climate change and knowledge of the damaging events that were about to unfold" (Oreskes and Conway, 2014, p. 10).

17 They note that the most enduring artistic text from the period is Kim Stanley Robinson's *Science in the Capital* trilogy (2004, 2005, 2007), which quickens the pace and effects of global warming to test various scenarios and human responses.

18 According to a recent ThinkProgress.org article, the Earth is heating faster than scientists predicted (Romm, 2014).

19 Toscano's is the first in what developed into an exchange between him and Jasper Bernes. Cf. Bernes, 2013, forthcoming; and Toscano, 2014.

20 Coulthard writes, "first, the practices are directly undertaken by the subjects of colonial oppression themselves and seek to produce an immediate power effect; second, they are undertaken in a way that indicates a loosening of internalized colonialism, which is itself a precondition for any meaningful change; third, they are prefigurative in the sense that they build the skills and social relationships (including those with the land) that are required within and among Indigenous communities to construct alternatives to the colonial relationship in the long run" (Coulthard, 2014, p. 166). The politics of the direct action are also briefly raised in Neil McArthur and Warren Cariou's short film *The Land of Oil and Water: Aboriginal Voices on the Oil Sands* (2010).

21 See Wilson, 2014; and Idle No More, 2013.

References

Barney, D. (forthcoming). Who We Are and What We Do: Canada as a Pipeline Nation. In Carlson, A., Szemand, I. and Wilson, S. (eds) *Petrocultures: Oil, Energy, Culture*. Montreal-Kingston: McGill-Queen's University Press.

Bernes, J. (2013) Logistics, Counter-logistics, and the Communist Prospect. *endnotes*. endnotes.org.uk/en/jasper-bernes-logistics-counterlogistics-and-the-communist-prospect.

Bernes, J. (forthcoming) The Belly of the Revolution: Agriculture, Energy, and the Future of Communism. In Bellamy, B.R. and Diamanti, J. (eds) *Materialism and the Critique of Energy*. Edmonton and Chicago, IL: MCMPrime Press.

Buell, F. (2012) A Short History of Oil Cultures: Or, the Marriage of Catastrophe and Exuberance. *Journal of American Studies*, 46, pp. 273–293.

Chen, A. (2014a) Pipeline Mapping (Blogpost). pipelinemapping.tumblr.com.

Chen, A. (2014b). Ann Chen: Identifying the Zeitgeist of Canada's Proposed Enbridge Northern Gateway Pipeline. *National Geographic*, October 27. voices.nationalgeographic.com/2014/10/17/ann-chen-identifying-the-zeitgeist-of-canadas-proposed-enbridge-northern-gateway-pipeline/.

Coulthard, G. (2014) *Red Skins White Masks: Rejecting the Colonial Politics of Recognition*. Minneapolis: University of Minnesota Press.

Diamanti, J. (2016) Three Theses on Energy and Capital. *Reviews in Cultural Theory*, 6(3), pp. 13–16.

Ferguson, E. (n.d.). *Line 9 Communities*. line9communities.com.

Foster, J.B. (1999) Marx's Theory of Metabolic Rift: Classical Foundations for Environmental Sociology. *American Journal of Sociology*, 105(2), pp. 366–405.

Greene, T. (2014). Protesters Square Off Against Kinder Morgan on the Mountain. *The Huffington Post*, November 25. www.huffingtonpost.ca/capt-trevor-greene/kinder-morgan-pipeline-protests_b_6216288.html?utm_hp_ref=enbridge.

Harvey, D. (2003) *The New Imperialism*. Oxford: Oxford University Press.

Huber, M.T. (2013) *Lifeblood: Oil, Freedom, and the Forces of Capital*. Minneapolis, MN: University of Minnesota Press.

Hui, S. (2014) SFU Students Organize Petition against Kinder Morgan Lawsuit Targeting Protesters (Blogpost), November 13. www.straight.com/blogra/770646/sfu-students-organize-petition-against-kinder-morgan-lawsuit-targeting-protesters.

Idle No More (n.d.) The Vision. www.idlenomore.ca/vision.

Idle No More (2013) Idle No More: How it Began. *The Province*, January 7. www.theprovince.com/news/bc/Idle+More+began/7780924/story.html.

Insko, J. (n.d.). Line 6B Citizens (Blogpost). grangehallpress.com/Enbridgeblog/.

Johnson, B. (2014) *Carbon Nation: Fossil Fuels in the Making of American Culture.* Lawrence, KS: University of Kansas Press.

LeMenager, S. (2014) *Living Oil: Petroleum in the American Century.* Oxford: Oxford University Press.

Malm, Andreas. (2016) *Fossil Capital: The Rise of Steam Power and the Roots of Global Warming*. London: Verso.

Marriot, J. and Minio-Paluello, M. (2013) *The Oil Road: Journeys from the Caspian Sea to the City of London*. London: Verso.

Marx, K. (1973) *The Grundrisse*. London: Penguin.

Mitchell, T. (2011) *Carbon Democracy.* New York: Verso.

O'Malley, M. (1976) *The Past and Future Land: An Account of the Berger Inquiry into the Proposed Mackenzie Valley Pipeline.* Toronto: Peter Martin Associates Limited.

Oreskes, N. and Conway, E.M. (2014) *The Collapse of Western Civilization: A View from the Future.* New York: Columbia University Press.

Petrocultures (2013) petrocultures.com/?page_id=10.

Romm, J. (2014) Earth is Heating Faster than We Realized, Making 2°C Limit for Global Warming More Urgent. *Climate Progress*, October 6. thinkprogress.org/climate/2014/10/06/3576154/2c-limit-global-warming/.

Simpson, M. (forthcoming) Lubricity: Smooth Oil's Political Frictions. In Carlson, A., Szeman, I. and Wilson, S. (eds) *Petrocultures: Oil, Energy, Culture.* Montreal-Kingston: McGill-Queen's University Press.

Solnit, R. (2003) The Annihilation of Space and Time. *New England Review*, 24(1), pp. 5–19.

Stanton, K. (2012) Looking Forward, Looking Back: The Canadian Truth and Reconciliation Commission and the Mackenzie Valley Pipeline Inquiry. *Canadian Journal of Law and Society*, 27(1), pp. 81–99.

Stern, P. (2007) Hunting for Hydrocarbons: Representations of Indigeneity in Reporting on the New Mackenzie Valley Gas Pipeline. *American Review of Canadian Studies*, 37(4), pp. 417–441.

Stoekl, A. (2014) Foreword. In Barrett, R. and Worden, D. (eds) *Oil Culture* (pp. xi–xiv). Minneapolis, MN: Minnesota University Press.

Szeman, I. (2007) System Failure: Oil, Futurity, and the Anticipation of Disaster. *South Atlantic Quarterly*, 106(4), pp. 805–823.

Szeman, I. (2013) How to Know About Oil: Energy Epistemologies and Political Futures. *Journal of Canadian Studies*, 47(3), pp. 145–168.

Toscano, A. (2011) Logistics and Opposition. *Mute*, August 9. www.metamute.org/editorial/articles/logistics-and-opposition#sdendnote8anc (accessed October 17, 2014).

Toscano, A. (2014) Lineaments of the Logistical State. *ViewPoint Magazine*, September 28. viewpointmag.com/2014/09/28/lineaments-of-the-logistical-state/.

Wilson, S. (2014) Gendering Oil: Tracing Western Petrosexual Relations. In Barrett, R. and Worden, D. (eds) *Oil Culture* (pp. 244–263). Minneapolis, MN: University of Minnesota Press.

9 Yugoslavism

History, temporality, and the search for alternative modes of political critique

Petra Rethmann

In May 2013 I visited Zagreb to attend the Second Balkan Forum. Every day from May 7–18, Zagreb's Youth Theater courtyard was filled with the sound of clinking coffee cups and voices as union activists, teachers, organized students, and others met to discuss themes spanning the range of feminism, LGBT activism, 21st-century imperialism, workers' struggles and co-ops, the demise of social welfare programs, international solidarity and the European debt crisis. Sponsored in part by the Rosa-Luxembourg Foundation, Heinrich-Böll Foundation, PEN Croatia, and Attac France, in 2012 the Forum had been created as a meeting point for activists from Poland, Albania, Bosnia, Serbia, Romania, Bulgaria, Slovenia and Croatia, with self-identified comrades from Italy, Spain, Germany and Greece attending as well. With the Greek debt crisis looming and critiques of European Union (EU) economic politics on the rise, every participant I met and to whom I talked spoke about the negative outfalls of privatization (denationalization and deregulation of industry, downscaling of government programs with the goal of increasing market efficiencies), stabilization (anti-inflation policies, hard budget constraints and a convertible currency), and liberalization (domestic market competition). In particular, in speaking from within (and about) what is commonly referred to as "the post-socialist condition" – a political reality and structure of feeling that indexes that there is no alternative – Eastern European activists expressed grave concerns about job losses, cutbacks in pay, benefits, and pensions, the corporatization of university education (Živkovič, 2015), and the general decline in public and collective forms of life. At the same time, they also spoke about the need for cosmopolitan democracy and transnational solidarity, even as they realized that these terms indicated more of a dream than a reality for the globe.

Central to many assessments and accounts that emerged at the meeting in Zagreb was what Peter, a young Serbian sociologist and activist,[1] expressed as the "politics of hope": a political attitude and stance that refuses to believe in the dictum that there is no alternative to direct attention to the fact that there should be an alternative. Peter located this politics concretely in Yugoslavism, an idea (elaborated below) that has been circulating in a number of forums and political groups. Yugoslavism evokes cosmopolitan federalism as a political form through which to democratize governance on a transnational scale. Peter's remarks reflected Yugoslavism's location in the context of a utopian imagination that asks critical analysts and political activists to think about what Elizabeth Povinelli (2011) has called an "otherwise" and Jennifer Wenzel (2009) an "otherwhile."

In thinking out loud about the possibilities for constitutional arrangements that recognize plurality, diversity, and fragmentation within a nationally disaggregated polity, Peter's remarks reflected Yugoslavism's connection of the need for social and economic solidarity to a more decentralized form of sovereignty. In essence, in rejecting the axiom that political communities should be grounded in national identities, Yugoslavism envisions a possible form of regional confederation and transnational democracy.

Yugoslavism very helpfully illustrates the links between past political practices, dreams of a different future, and new ways of imagining the local and the global, which express the complex and productive links between time and globalization that are the focus of this book. In this chapter I am also interested in understanding how and why Yugoslavism continues to constitute a political space of attraction for Eastern European subjects. In following others (Štiks and Horvat, 2015), I suggest that it might provide the left in the region with a stronger sense of community, and the ability to link its struggles in more effective ways. Yet for this transnational vision to emerge, a number of obstacles relating to the politics of temporality and time have to be cleared out of the way, including some contemporary historiographical and analytical conventions that may preclude analysts and activists from grasping Yugoslavism as an alternative mode of political analysis and thinking. I am not suggesting that in this present Yugoslavism constitutes an answer for political desires that wish to reconcile, say, cultural singularity and multiplicity, or wish to establish social mutuality and equality, but together with Peter and others I argue that contemporary social and critical thinking should not so easily dismiss previous attempts to imagine (if imperfectly implemented) alternative political worlds.

I begin this analysis with a brief narrative of Yugoslavism as one site of difference within the context of an alleged utopia of socialist modernization and an integral part of the non-aligned movement (NAM) (Prashad, 2007). In thus beginning in the middle of Yugoslavism's story, I skip over its 19th-century history as one possible route to promote Serbo-Croatian unity, as well as a political practice supposed to transcend the religious, national, linguistic, and historical differences that separated south Slavic cultural groups and nations (Trgovčević, 2003; Todorova, 2009; Pavković, 2003). I also skip over the social activism and thought of 19th-century Serbian socialist Svetozar Marković (Grubačič, 2010, pp. 162–164), not because I wish to dismiss its underlying impetus to address economic disparity and ethnic discrimination, but because Peter's comments can be framed by his contribution first and foremost in terms of political relations. While from a perspective of today it may appear more than obvious that Yugoslavism indexes nothing more than a failed political project (Bieber, 2013), cultural and parochial idea (Drapac, 2010), or outdated curio from the archives of left history (Djokič, 2013; Djilas, 2013), for some subjects at least it constitutes a vital legacy from which to imagine another future. In taking this understanding seriously instead of quickly dismissing it out of hand, this chapter accepts the suggestion that Peter expressed as an invitation to think about Yugoslavism – including our stories about and methods for approaching its own temporality and time – as a possible source of inspiration for political imaginations and thinking.

Once I have established some of the more hopeful aspects and dimensions of Yugoslavism as a practice and idea, I shift the analysis to more critical inquiries into the politics of temporality and time. In keeping in step with the expression of Yugoslavism as a concept of hope, I think about dimensions of temporality associated with the term as non-linear, multidirectional, and simultaneous in a way that opens it up to the possibility of conceiving of it as a present force. Here historian Reinhart Koselleck's (2004, p. 95) notion of "the contemporaneity of the noncontemporaneous" that signals how this now – or what he calls a temporal refraction – contains a diversity of temporal strata acts as my guide, as does the political utopianism of unorthodox Marxist thinker Ernst Bloch. In building in particular on Bloch's major work in three volumes, *The Principle of Hope*, I examine Yugoslavism's double focus as affirmation and critique. In looking at the diagnostic presumptions of some of our temporal conventions, including historical linearity and post-socialist and left-wing nostalgia, I emphasize hope's capacity for what Viktor Shklovsky (2002) has characterized as estrangement (*ostranenie*) – the ability to render unfamiliar the all-too-recognizable contours of present configurations of time and the temporal meanings to which we have become habituated. My goal here is to distance analysis from the present order, to suspend and momentarily disable those epistemologies and habits of thinking that keep subjects locked in narrow orbits of social possibilities and options. In then directing this chapter's analytical gaze toward hope's provocative function, I turn my attention to the role of dreaming and the untimely as markers of possibility of different futures. In thus seeking to grasp those elements that mark Yugoslavism as a de-reifying and affirming technique, I also investigate Yugoslavism as what Bloch (1995) calls the "real – possible" – a project of hope, to be sure, but one instantiated in a concrete narrative of politics and time that inevitably constrains the temporal possibilities explored below.

A caveat is in order before the analysis begins: in this chapter globalization does not appear in the form of corporate market economy, the spread of Western liberal democracy (Fukuyama, 1989), or a worldwide political advance to a global era (Hardt and Negri, 2001). Neither does it appear as a congeries of local/global interactions, or an emphasis on – as anthropologists are prone to do – the specificity of cultural productions in relation to globalization. Rather, it appears as a concern and desire: as a post-socialist resident dream connected to other dreams of equality and freedom – to, for example, the internationalist post-World War II dream of Bandung, the economic autonomy and political self-management vision of Aimé Césaire, the socialism of Michael Manley. Of course – Bandung, Césaire, and Manley – these are names that have become symbols of what was once thought of as possible: national sovereignty and world peace. These are also names connected to a dream space that since the 1970s has been slowly coming apart. Usually, this dream space is now exhibited in dioramas, in museums behind glass, visible but dusty. Yet critiques of globalization as a form of corporate consolidation, standardization of world markets, and deference of national governments to transnational business demands require more than – as suggested in the anti-/alter-globalization movement (AGM) – visions of networks, rhizomes, assemblages, webs, and braids (Rethmann, 2013; Graeber, 2009; Khasnabish, 2013). They also require temporal bridges that assist analysts-activists in understanding the dream of Yugoslavism as a connective one – an internationalism of lost possibilities from which political subjects might learn: not as a legacy to preserve but as a tool to use.

Yugoslavism

At the Yalta conference that took place on February 4–11, 1945, between US President Franklin D. Roosevelt, British Prime Minister Winston Churchill, and Soviet Premier Joseph Stalin, a number of important decisions regarding the future of a postwar world were being made: Germany was to be divided into four occupied zones (*Besatzungszonen*) administered by US, French, British, and Soviet forces, and – following Japan's surrender – the Soviets were granted a sphere of political influence in Manchuria, including a share in the operations of the Manchurian railroads and the Kuril Islands. The American and British governments agreed that future governments of the Eastern European nations should be "friendly" to the Soviet regime, while the Soviets pledged to allow free elections in all territories liberated from Nazi Germany. In addition, negotiators also released a communiqué on Poland, allowing for the inclusion of communists in its postwar national government. As subsequently the Yalta scheme began to take shape in the world, it also produced a binary political logic that argued that nation-states had to line up behind one or the other of the two superpowers. Globally, neutrality or non-alignment simply no longer seemed to be an option.

In Yugoslavia the world-historical transformations that happened in the wake of Yalta articulated themselves perhaps most sharply in 1948 – the year when Yugoslavia was expelled from the Communist Information Bureau (Cominform). Created in September 1947 at a meeting in the Polish town of Szklarska Poreba and run by Moscow's propaganda chief Andrei Zhdanov, the Cominform indexed the means by which the Soviet Union communicated its political line to Eastern European leaders in order to direct their policies. At around the same time, Yugoslavia's leader Josip Broz Tito – a Croatian trade unionist who had fought in World War I, participated in the Bolshevik Revolution, fought in the Red Army, joined the Communist Party of

Yugoslavia, led the Dimitrov Battalion in the Spanish Civil War, returned to Yugoslavia to fight a partisan-driven guerrilla war against the fascists, and in March 1945 became the premier of the federation of the socialist republics of Bosnia-Herzegovina, Croatia, Macedonia, Montenegro, Serbia, and Slovenia – began to clash with Soviet leader Stalin. Concretely, Tito's Yugoslavia supported communist revolutionaries in neighboring Greece whom Stalin considered as falling within the American and British sphere of influence. In refusing to be beholden to a foreign power, Tito also actively began to work on the institutionalization of a Balkan federation that would merge the Socialist Federal Republic of Yugoslavia (SFR Yugoslavia) with Bulgaria. Angered by these signs of Yugoslavia's political independence, the Soviets sought to isolate Yugoslavia, including its formal exclusion from the socialist camp. As one response, Yugoslavia began to seek political bonds with decolonizing nations that expressed interest in a democratic and non-superpower federation, including India and Egypt.

In mid-July 1956 – one year after the 1955 Asia-Africa conference in Bandung – India's Prime Minister Jawaharlal Nehru, Egypt's President Gamal Abdel Nasser, and Tito gathered on the Yugoslav island of Brijuni to discuss the fallout from the Cold War and their own vision for a non-aligned force in opposition to the Yalta logic. In reviewing international developments and in also assessing the effects of Bandung, they began to formulate the agenda of what in 1961 in Belgrade would be formalized as the NAM. Although the national and international politics of those nations – Mali, Burma, Cambodia, Ethiopia, Ghana, Yugoslavia, India, among others – that aligned themselves formally in the NAM were deeply uneven, they came together in a desire to opt out of Zhdanov's two-camp theory and the globe's larger, bipolar conflict. In doing so, they emphasized the need for the formulation of alternative political agendas, including non-superpower democracy on a supranational scale. For example, in deploying a political language that emphasized terms such as "non-alignment," "peaceful co-existence," and "democratic cooperation," they argued for a politically internationalist vision that involved mutuality and the recognition of multiplicity and diversity. In casting international relations in terms of political ethics rather than in terms of national interests or power, for many – in particular Third World – nations the NAM stood as a beacon of freedom, entailing the twofold promise of global democracy and solidarity.

In identifying a global democratic federalism as the most effective form for Third World and other "minority" nations through which to achieve formal political equality and independence, the NAM situated itself within a politically utopian tradition at once strategic and untimely (Goswami, 2012). In conceptualizing international and democratic federalism in terms of mutuality and complementarity instead of servility, and in understanding it as a means for creating a political community and not simply as an instrument for gaining a political voice and power, the NAM posed a challenge to a politics of instrumental calculation. In fact, in considering mutuality as a starting point for a less hierarchical world, the NAM argued for the reorganization of the United Nations (UN), one built less on a territorial and imperial nationalist logic and more on political emancipation and reciprocal enrichment. Lurking behind its efforts and imaginations was a refusal to conflate popular sovereignty with territorial sovereignty, even as territorial sovereignty became the normalized and institutional form for autonomy and independence.

Examining the left past

To all appearances, the fragment narrated above constitutes just one more example of what David Scott (2004) has called a "future past": an historically superseded and obsolete political mode that can no longer animate utopian imaginations. In particular, in light of the terrible violence that followed on the heels of the 1990s break up of SFR Yugoslavia, images of Yugoslavia as ethnically essentialist, uncivilized, archaic, or simply crazy began to shape Northern and Western imaginations even as the political conditions that allegedly gave rise to them disappeared (Kaplan, 2005). In turn, reductive understanding of Yugoslavism as a failed form of democratic federalism started to calcify, and its open and cosmopolitan aspects simply shriveled and fell away (Todorova, 2013). As one consequence, Yugoslavism signified nothing more than, in Manu Goswami's (2012, p. 1462) words, a "holding operation" against the inevitable move from multi-national federation to singular nation, with the post-Yugoslav nation-form indexing a purportedly primordial form of belonging. The idea and practice of a cosmopolitan, multi-cultural, and federalist state equaled catastrophe and failure. Certainly the evocation of Yugoslavism as an expression of hope can be criticized, but it is important nevertheless to direct attention to how some historiographical assessments have petrified and frustrated a critical engagement with the valuable utopian potential of Yugoslavism.

Consider, for example, the recent stand-off between Marxism and anarchism (Graeber, 2013; Ross, 2015). In what has become a well-known and often repeated taxonomy, Marxism and anarchism are posed as competing models of left theory, analysis, and practice – with Marxism posited as a model that centers on economic domination, and anarchism as one that focuses on political domination. One historical way in which this taxonomy articulates itself is in the production of a political origin narrative that marks the January 1, 1994 uprising of Mexico's Ejército Zapatista de Liberación Nacional (Zapatista National Liberation Army, EZLN) in the Mexican state of Chiapas as a new beginning for a global left (Graeber, 2009; Juris and Khasnabish, 2013). In fact, in positing the Zapatistas "as the catalyst for a profound [political] reimagining on a transnational scale of the possibilities for and paths toward radical social change" (Khasnabish, 2013, p. 68), AGM activists and analysts have chartered their own foundational and genealogical origin. In particular by employing the botanical metaphor of the Deleuzian rhizome – a subterranean plant that produces shoots, tubers, bulbs, and nodes (but does not tend to develop deep roots) – as one of alter-globalization movements' defining metaphors, activist-analysts emphasize the importance of a horizontal instead of a vertical politics – with the latter being defined by The Party. In thus stressing the spatiality instead of temporality of an innovative and critical politics, AGM analyses build on and advocate political imaginations of enlargement and extension, instead of tracing untimely processes and lines of liberatory descent. The problem is not that implicitly AGM analyses point to some of the limits of Peter's evocation, but that they can keep activists and analysts from creative appropriations of some of his insights.

In the case that concerns me here, an affirmative understanding of Yugoslavism as an articulation of hope is thus made difficult by some of the left's own historical practices, including some of its own familiar periodizing models and classificatory schemes. In particular, in conceiving of the relationship between a socialist- and internationalist-oriented past and rhizomatic present in terms of eras in succession, AGM analysts and activists tend to run up against the question of why one would bother with the past. The problem is then not only that recent placements of anarchism within the context of a temporally linear logic are too dismissive of Yugoslavism as an historical past, but that they block access to a full and rich engagement with this past. It is for this reason that recently critical analysts have urged others to pay attention to the relationship between the tenses, including the relationship between history, presentism, and futurity (just think of the burgeoning interest in Walter Benjamin's innovative constellations of time). I will return to this subject at the end of this piece. For now, suffice to say that what may also be needed is what Kathi Weeks (2011, pp. 171–174) has called a "time movement": not as a subsidiary to contemporary social movements but as a shift in left historiography to grasp the transformative potential inherent in bringing past and present into communication so that they can – it is hoped – distort and remake one another.

Memory and left-wing nostalgia

As has been commented on so many times, in Eastern European and post-socialist analyses a veritable concern with the forms and politics of memory emerged in the years after the 1989–91 demise of the state-socialist project in Eastern Europe and the Soviet Union (Blacker et al., 2013). In particular in conjunction with the memory boom that took hold of the social sciences and humanities in the 1980s and 1990s (Huyssen, 2003; Sturken, 2007), anthropologists and other critical analysts began to look at the ways in which Eastern European actors became invested in remembering, memorializing, narrating, or inventing the past. In being especially concerned with the ways in which Eastern European subjects articulated affective attachments in relation to experiences of loss, including sadness, mourning, and nostalgia, analysts troubled a number of habitual temporal assumptions, especially those that assumed stable distinctions between the present and past. Yet in uncritically appropriating disciplinary history's understanding of time as a neutral medium within which history takes place, rather than looking at time as a productive historical force of its own, anthropologists have also tended to situate an evocation such as Peter's within a politics of time that can only understand Yugoslavism as a symptom (Boyer, 2006): that is, as just one more expression of "Yugonostalgia," "Titostalgia" (Velinkonja, 2015), political injury and social fear, or an affective displacement of a once-existent promise of a better life. Haunted by a past that overwhelms the present, in this sense Yugoslavism marks nothing more than a symptomatic return of a time that has already been experienced and lived.

Resonances of Yugoslavism as a nostalgic symptom can also be found in left readings of the socialist past that acknowledge its historical value in terms of melancholia (Scribner, 2005) or restorative consolation (Boym, 2002). In exhibiting a "wounded attachment" (Brown, 1995) to a past that supposedly can no longer be grasped in terms of freedom, a utopian future, political possibility, and justice, the left then develops its own mode of nostalgia (Benjamin, 1999; Brown, 2003): one that mourns its own often idealized past, its now superseded forms of organizing, and political experiences and ways of being. In indexing an affective relationship to time that tends to be more backward looking than anticipatory, more attached to its own (political) narcissism and sorrow than to the possibility of its overcoming, it constitutes the left subject as one of ressentiment – as one marked by a pessimism of the will and cynical reason. The task, however, is not to forget the past or to disavow the lingering impact of the past on the present and future – the reality of multiple kinds of pain and suffering should be acknowledged and not denied – but to achieve a relationship to the past that could be more enabling of a different future.

One way to achieve such a relationship is to recognize the evocation of Yugoslavism as a site of hope not as an epistemological proposition but as a performative and ontological intervention. In other words, it is not really a question of if nostalgia can help analysts or activists to become more interested in this world (Flatley, 2008), or if analysts and activists believe in or wish for Yugoslavia's return. What is at stake is not the facticity and past-ness of Peter's evocation, but its critical and affirmative impact on the political imagination. In offering a different temporal perspective, the framing of Yugoslavism as a mode of political hope challenges this present's political order as the crowning achievement of an inevitable historical development and, in turn, makes possible a relativizing of the present. In seeking to estrange listeners and readers from an often all-absorbing familiarity with judgments about the historical past, it seeks to provide them with a standpoint from which to assess this present critically. As a performative provocation, it provides a glimmer of a different world, hoping to suspend or at least momentarily disable those habits of temporal thinking that keep us locked within a narrow orbit of political possibilities and options – not to fix but to evoke alternatives.

Dreaming and dreams

In this current political climate, it is easy to dismiss Peter's evocation of Yugoslavism and the analytical trajectory I have offered here so far in terms of a "politics of mereness" (Berdahl, 1999, p. 193). From such a perspective, Peter's evocation is "merely hopeful," "merely dreamy," or "merely idealistic": in other words, it appears as fantasy and just talk. Such customary political logic makes it easy to write off Yugoslavism as a distraction and diversion from more realist struggles for social change. Rather than waste (analytical) energy and time on the impracticality and impossibility of Yugoslavism as a utopian imagination and political dream, analysts and activists should set their sights on politically realistic and feasible goals. In addition, the pressures of getting by in hard times (Ehrenreich, 2011; Kideckel, 2008) tend, as Robin Kelley (2002) has noted, not to be generative of utopian imaginations. In fact, consumed by this now, the possibilities of alternatives to the ever more reified structures of this supposedly post-ideological moment and neoliberal life appear increasingly remote. As cultural critic Tom Moylan (2000, p. 103) describes it, "the leaner and meaner [...] 1980s and 1990s [created a world of] anti-utopian deprivation rather than utopian achievement."

As one response, in the last few years political and critical analysts have started to reclaim the political significance of the utopian imagination, including the social power of dreams (Duncombe, 2007; Rethmann, 2004). In arguing that imaginations of alternative futures often entail glimpses across temporal divides, they have paid special attention to Marxist philosopher Ernst Bloch's conception of "nonsynchronous contradictions" – whereby past forms of life persist in a disjunctive present – and his contention that the capacity for thinking and wanting the new tends to appear in dreams, particularly daydreams. In situating the political importance of the dream as a polemical defense against a Marxist anti-utopianism (see also Rancière, 1989), as well as a means to provoke the political imagination, Bloch challenges assumptions that dismiss the political dream as trivial and irrelevant. In conceptualizing dreams as "anticipating elements, [as] a component of reality itself," he questions the view that what counts as real can be isolated as fact from a process of becoming. After all, a "real realism" demands the recognition that the future is born in every present, and that what it will become is not yet decided – thus grasping the present in relation to both its genealogies and its fronts. Dreams indulge desires for different futures, experiment with ways to fulfill them, and may enjoy their imagined satisfaction. While they are not utopian imaginations, in them analysts may well catch a glimpse of the not-yet that animates utopian imaginations: as a nascent expression of a desire for new and better forms of life, even if only in limited form.

In light of the fact that Bloch constitutes the dream as one trace of an alternative mode of political desire, it may be worth pointing out that the dream he speaks about is not the Freudian night dream that constitutes a particular form of remembering. Neither is it a ghostly image that haunts a waking present. Rather, as entities that mark possible processes of emergence, dreams direct analysts' attention toward the latent power congealed in political projects that are "not yet conscious" or "not yet realized." In signifying the world as "unenclosed" (Bloch, 1995, p. 246) and the future as open, they anchor possibilities of what might become. In order to grasp these possibilities, then, analysts must not only understand a dream's emergence from and attachment to the past but also its leading edges. Or, as Kathi Weeks (2011, p. 189) puts it, "everything real has not only a history, but also a horizon."

One of the reasons why Bloch's understanding of dreams is so important for critical analysts is that it helps them to recognize the significance of a speculative imagination and assists them in putting it into practice. Rather than insist that Peter's evocation is "merely hopeful" or "merely dreamy," it allows them to conceptualize dreams as a particular mode of temporal reasoning – thinking backward and thinking forward, thinking the fullness of any one moment in time. Understanding Peter's evocation of Yugoslavism as a (however fragmentary) dream does not signify him or his words as "merely dreamy": instead, in blurring boundaries between then and now, reality and fantasy, experience and imagination, it serves as a relay point between history, this present, and an untimely conviction that a better world is possible and yet to come.

Untimeliness and hope

Hope was the word Peter used at the Second Balkan Forum when he expressed the vision of Yugoslavism as one possible dream for a democratic federalism and future. In refusing a triumphant neoliberal and historical logic that asserts that there is no alternative or a narrow realism that demands that analysts and activists become practical, realistic, and outgrow utopia, this vision imagines a more expansive mode of political practice and the political imagination. As Bloch (1995, pp. 1–3) describes it, "the emotion of hope goes out of itself, makes people broad instead of confining them." What seems crucial in Peter's evocation of Yugoslavism as a politics of hope is that it extends Yugoslavism beyond a focus on its future past, or mode of failure or ressentiment. Yugoslavism is a political project, a project of hope, and thus does not only seek to counteract a politics of cynicism, pessimism, or fear, but also encourages another approach to time. Rather than situate Yugoslavism in an historical imagination of linear progress, of ending and failure, Peter's hope and project of hope asks analysts to examine his evocation from the perspective of a temporally uneven and densely layered present. From such a perspective, too, Yugoslavism appears as one optic that refracts politics as a number of temporal positions that either create or do not create the conditions in which not only unrealized pasts but imagined futures may emerge.

As if to respond to such challenges, recently historians, anthropologists, and critical analysts (Ross, 2009; Wenzel, 2009; Wilder, 2015; Brown, 2005) have called attention to the untimely – that is, to the social fact that the historical present is not always identical with itself – qualities of political possibilities and time. Their mediations on historical ruptures, blurred tenses, and possible futures entail processes of temporal confusion or illumination when conventional differentiations between past, present, and future no longer obtain. While in this article I have emphasized the untimeliness of hope as an analytically liberating and positive dimension, it should be noted that untimely practices can serve both politically progressive and conservative ends. It is in this sense that hope should not be confused with either idealism or futurism. Here a number of dreams of change that entail no heterodox conceptions of temporality come to mind: from a neoconservative ideal of national solidarity anchored by the family to neoliberalism's post-political vision of a world made free by the unhindered reign of the market. While these dreams may contain some utopian longings, they suggest for the most part allegedly better versions of the present rather than visions of a different world. Equally, the utopian imagination needs to be distinguished from futuristic thinking. Rather than imagine the future in terms of a predictable evolution from the present, the utopian imagination approaches time as contingent, with possibilities for significant ruptures.

Peter's evocation of Yugoslavism as hope is not a facile project: it is not an easy thing to be in time in this way, to straddle past, present, and future, to nurture both critical and affirmative investments in the lived present as it emerges from the past and its possible futures. Certainly, this dream of Yugoslavism has its limitations. In fact, one of the most fundamental paradoxes of Yugoslavism is that instantiating it in a narrative, a call, an evocation, utopian hope is at once awakened and delimited. Perhaps the most we can hope is that Yugoslavism can still animate imaginations and produce enticement: an invitation to think "otherwise" to produce "otherwhiles."

There is perhaps one point to add to this conclusion. In emphasizing hope's dual function of estrangement and affirmation, this analysis has not – as analysts usually do – provided an explicit evaluation of Peter's evocation. Instead, in following Peter's expression of Yugoslavism as hope and utopian imagination, it has asked readers to engage in a practice of political analysis and participate in a process of critical reflection. Because this vision of Yugoslavism is not as an empirical reality but rather a political possibility, it provokes us, as E.P. Thompson (1976, p. 97) put it, to "want [to] desire better, desire more, and above all desire in a different way," to affirm the utopian imagination less as an artifact than a practice, less for its content than what it asks analysts and activists to do – as an imperative to imagine alternatives.

Note

1 Peter's remarks were similar to many such comments about the politics of hope and Yugoslavism that have been expressed at this event and elsewhere. They echo another argument regarding Yugoslavism – an argument that actually draws on the

19th-century history of the idea (Rethmann, 2015). My references to Peter in the remainder of this chapter are figurative, and a shorthand for this larger conversation.

References

Balser, M., Harvey, F. and Glenn, M. (eds) (2004) *In the Way of Development: Indigenous Peoples, Life Projects, and Globalization: A Dream of Democracy in the Russian Far East* (pp. 256–278). London: Zed Books.

Benjamin, W. (1999) Left-Wing Melancholia. In Jennings, M.W., Eiland, H. and Smith, G. (eds) *Walter Benjamin: Selected Writings, Vol. 2, 1927–1934* (pp. 423–427). Cambridge, MA: The Belknapp Press of Harvard University Press.

Berdahl, D. (1999) (N)ostalgia for the Present: Memory, Longing, and East German Things. *Ethnos*, 64(2), pp. 192–211.

Bieber, F. (2013) Federalizing the Federation: the Failure of the Yugoslav Experiment. In Loughlin, J., John, K. and Wilfried, S. (eds) *Routledge Handbook of Regionalism and Federalism* (pp. 513–526). New York: Routledge.

Blacker, U., Etkind, A. and Fedor, J. (2013) *Memory and Theory in Eastern Europe.* New York: Palgrave Macmillan.

Bloch, E. (1977) Nonsynchronism and the Obligations to its Dialectics. *New German Critique*, 11, pp. 22–38.

Bloch, E. (1995) *The Principle of Hope 3 vols.* (N. Plaice, S. Blum and P. Knight, Trans). Cambridge, MA: Massachusetts Institute of Technology Press.

Boyer, D. (2006) Ostalgie and the Politics of the Future in Eastern Germany. *Public Culture*, 18(2), pp. 361–381.

Boym, S. (2002) *The Future of Nostalgia*. New York: Basic Books.

Božović, M. (2013) Zenit Rising: Return to a Balkan Avant-Garde. In Gorup, R. (ed.), *After Yugoslavia: the Cultural Spaces of a Vanished Land* (pp. 135–156). Stanford, CA: Stanford University Press.

Brown, W. (1995) Wounded Attachments. In *States of Injury: Power and Freedom in Late Modernity* (pp. 52–76). Princeton, NJ: Princeton University Press.

Brown, W. (2003) Resisting Left Melancholia. In Eng, D.L. and Kazanjian, D. (eds) *Loss* (pp. 458–465). Berkeley, CA: University of California Press.

Brown, W. (2005) Untimeliness and Punctuality. In *Edgework: Critical Essays on Knowledge and Politics* (pp. 1–17). Princeton, NJ: Princeton University Press.

Djilas, A. (2003) Funeral Oration for Yugoslavia: An Imaginary Dialogue with Western Friends. In Djokič, D. (ed.), *Yugoslavism: Histories of a Failed Idea* (pp. 317–333). London: Hurst & Company.

Djokič, D. (2013) The Past as Future: Post-Yugoslav Space in the Early Twenty-First Century. In Gorup, R. (ed.), *After Yugoslavia: The Cultural Spaces of a Vanished Land* (pp. 55–87). Stanford, CA: Stanford University Press.

Drapac, V. (2010) *Construction Yugoslavia: a Transnational History.* London: Macmillan.

Duncombe, S. (2007) *Re-Imagining Progressive Politics in an Age of Fantasy.* New York: The New York Press.

Ehrenreich, B. (2008) *Affective Mapping: Melancholia and the Politics of Modernism.* Cambridge, MA: Harvard University Press.

Ehrenreich, B. (2011) *Nickel and Dimed: On (Not) Getting by in America*. New York: Picador.

Eng, D.L. and Kazanjian, D. (eds) (2003) *Loss: Resisting Left Melancholia* (pp. 458–465). Berkeley: University of California Press.

Flatley, J. (2008) *Affective Mapping: Melancholia and the Politics of Modernism*. Cambridge, MA: Harvard University Press.

Fukuyama, F. (1989) The End of History. *National Interest*, 16, pp. 3–18.

Gorup, R. (ed.) (2013) *After Yugoslavia: the Cultural Spaces of a Vanished Land: My Yugoslavia* (pp. 23–37). Stanford, CA: Stanford University Press.

Goswami, M. (2012) Imaginary Futures and Colonial Internationalisms. *American Historical Review*, 117(5), pp. 1461–1485.

Graeber, D. (2009) *Direct Action: an Ethnography*. Oakland, CA: AK Press.

Graeber, D. (2013) *The Democratic Project: A History, a Crisis, a Movement*. New York: Spiegel and Grau.

Grubačič, A. (2010) *Don't Mourn, Balkanize! Essays after Yugoslavia*. Oakland, CA: PM Press.

Hardt, M. and Negri, A. (2001) *Empire*. Cambridge, MA: Harvard University Press.

Huyssen, A. (2003) *Present Pasts: Urban Palimpsests and the Politics of Memory*. Stanford: Stanford University Press.

Juris, J. and Khasnabish, A. (2013) Introduction: Ethnography and Activism within Networked Spaces of Transnational Encounter. In Juris, J.S. and Khasnabish, A. (eds) *Insurgent Encounters: Transnational Activism, Ethnography, and the Political* (pp. 1–36). Durham, NC: Duke University Press.

Kaplan, R. (2005) *Balkan Ghosts: A Journey through History*. New York: Picador.

Kelley, R.D.G. (2002) *Freedom Dreams: the Black Radical Imagination*. Boston, MA: Beacon.

Khasnabish, A. (2013) Tracing the Zapatista Rhizome, or, the Ethnography of a Transnationalized Political Imagination. In Juris, J.S. and Khasnabish, A. (eds) *Insurgent Encounters: Transnational Activism, Ethnography, and the Political* (pp. 66–88). Durham, NC: Duke University Press.

Kideckel, D. (2008) *Getting in Post-socialist Romania: Labor, the Body, and Working-Class Culture*. Bloomington, IN: Indiana University Press.

Koselleck, R. (2004) *Futures Past: On the Semantics of Historical Time*. New York: Columbia University Press.

Moylan, T. (2000) *Scraps of the Untainted Sky: Science Fiction, Utopia, Dystopia*. Boulder, CO: Westview Press.

Pavković, A. (2003) Yugoslavism's Last Stand: a Utopia of Serb Intellectuals. In Djokič, D. (ed.), *Yugoslavism: Histories of a Failed Idea* (pp. 252–267). London: Hurst & Company.

Povinelli, E. (2011) *Economies of Abandonment: Social Belonging and Endurance in Late Liberalism*. Durham, NC: Duke University Press.

Prashad, V. (2007) *The Darker Nations: a People's History of the Third World*. New York: The New Press.

Rancière, J. (1989) *Nights of Labor: The Workers' Dream in Nineteenth-Century France*. Philadelphia, PA: Temple University Press.

Rethmann, P. (2004) A Dream of Democracy in the Russian Far East. In Blaser, M. and Feit, H. (eds) *In the Way of Development: Indigenous Peoples, Life Projects, and Globalization* (pp. 256–278). London: Zed Books.

Rethmann, P. (2013) Imagining Political Possibility in an Age of Late Liberalism and Cynical Reason. *Reviews in Anthropology*, 42(2), pp. 226–242.

Rethmann, P. (2015) Internationalism, Temporality, and Hope: a View from Eastern Europe and the Left. *Social Anthropology*, 23(4), pp. 480–491.

Rockhill, G. and Watts, P. (eds) (2009) *Jacques Rancière: History, Aesthetics, Politics: Historicizing Untimeliness* (pp. 15–30). Durham, NC: Duke University Press.

Ross, K. (2009) Historicizing Untimeliness. In Rockhill, G. and Watts, P. (eds) *Jacques Ranciere: History, Politics, Aesthetics* (pp. 15–29). Durham, NC: Duke University Press.

Ross, K. (2015) The Paris Commune and the Literature of the North. *Critical Inquiry*, 41(2), pp. 269–288.

Scott, D. (2004) *Conscripts of Modernity: the Tragedy of Colonial Enlightenment*. Durham, NC: Duke University Press.

Scribner, C. (2005) *Requiem for Communism*. Cambridge, MA: Massachusetts Institute of Technology Press.

Shklovsky, V. (2002) *Third Factory*. Chicago, IL: Dalkey Archive Press.

Štiks, I. and Horvat, S. (eds) (2015) *Welcome to the Desert of Post-Socialism: Radical Politics after Yugoslavia: Introduction: Radical Politics in the Desert of Transition* (pp. 1–20). London: Verso.

Sturken, M. (2007) *Tourists of History: Memory, Kitsch, and Consumerism from Oklahoma City to Ground Zero*. Durham, NC: Duke University Press.

Thompson, E.P. (1976) Romanticism, Moralism, and Utopianism: The Case of William Morris. *New Left Review*, 99, pp. 83–111.

Todorova, M. (2009) *Imagining the Balkans*. Oxford: Oxford University Press.

Todorova, M. (2013) My Yugoslavia. In Gorup, R. (ed.), *After Yugoslavia: The Cultural Spaces of a Vanished Land* (pp. 23–37). Stanford, CA: Stanford University Press.

Trgovčević, L. (2003) South Slav Intellectuals and the Creation of Yugoslavia. In Djokič, D. (ed.), *Yugoslavism: Histories of a Failed Idea* (pp. 222–237). London: Hurst & Company.

Velinkonja, M. (2015) Mapping Nostalgia for Tito: From Commemoration to Activism. In Horvat, S. and Štiks, I. (eds) *Welcome to the Desert of Post-Socialism: Radical Politics after Yugoslavia* (pp. 173–198). London: Verso.

Weeks, K. (2011) *The Problems with Work: Feminism, Marxism, Anti-work Politics, and Postwork Imaginaries*. Durham, NC: Duke University Press.

Wenzel, J. (2009) *Bulletproof: Afterlives of Anticolonial Prophecy in South Africa and Beyond*. Chicago, IL: University of Chicago Press.

Wilder, G. (2015) *Freedom Time: Negritude, Decolonization, and the Future of the World*. Durham, NC: Duke University Press.

Živkovič, A. (2015) From the Market … to the Market: The Debt Economy after Yugoslavia. South Slav Intellectuals and the Creation of Yugoslavia. In Horvat, S. and Štiks, I. (eds) *Welcome to the Desert of Post-Socialism: Radical Politics after Yugoslavia* (pp. 45–65). London: Verso.

10 Living the rhythms of the global

Time, globalization discourse, and rhythm-analysis

Adam Barrows

Globalization demands the synchronized coordination of multiple and diverse rhythms, from the volatility of global stock markets to the variability of weather patterns, from the biorhythms of seasonal agricultural workers to the flex-times of corporate culture. Time may be conceptualized in a variety of ways but it is lived all over the globe as a negotiation of multiple and often competing rhythms. These necessarily diverse rhythmic patterns are uneasily harmonized by global processes that protect and privilege the economic, cultural, and political foundations of some rhythms at the expense of others. Conceptual models of "time under globalization," reflected in everything from colloquial expressions like "the world is getting so much faster," to more sophisticated theories of risk and volatility, can themselves be understood as globalizing assertions of homogeneity in the face of what is clearly a more protean reality. Despite suggestions that global temporal homogeneity and rhythmic uniformity are accomplished facts, globalization might more productively be understood as a perpetual negotiation between a host of competing and even mutually exclusive rhythmic patterns that constitute its global reality at any given moment and are thus always poised, if not empowered, to alter that reality.

Thinking rhythmically about the relationship between time and globalization means thinking not only about the rhythmic experiences of differently empowered human agents around the world but also about the natural rhythms of the globe itself, from global climate change to the rhythmic patterns of the non-human species with which human lives and economies are invariably intertwined and interdependent. Studying the relations between local temporal patterns of the smallest of life forms and larger-scale generational rhythms of time has typically not been a project of the humanities or social sciences, even as the intellectual disciplines that make up those fields have increasingly become attentive to the political imperatives of eco-criticism. At the forefront, however, of the most promising and politically engaged ecological research of the 21st century is a conception of the relationship between the local and the global that is not only inescapably rhythmic but is also attuned to the non-hierarchical nesting of smaller rhythmic patterns within larger ones, such that the local is understood as constitutive and enabling of the global. Ecological resilience and adaptation studies suggest, provocatively for the humanities, the influence of seemingly microscopic temporal and spatial patterns within the tenuous whole that makes up a larger ecosystem. Appropriately for the interdisciplinary volume in which this chapter is being published, I am suggesting an eco-political approach to the subject of time and globalization which draws as much upon an under-explored theoretical strain within the social sciences (namely the late work on rhythm-analysis by the Marxist sociologist Henri Lefebvre), as it does upon a body of work in the natural sciences which, as far as I am aware, has had no impact whatsoever on globalization discourse in the humanities.

This is not to say that the humanities, in order to intervene meaningfully within globalization studies, must simply borrow the paradigms and practices of the natural sciences. There is perhaps no better form of mediation between the intimate temporal rhythms of local experience and the demands of global rhythmic regimes than narrative, the telling of stories. Globalization is itself as much a set of processes for reshaping local economies and rhythms according to dominant models as it is a compelling story about those processes. Fictional narration is especially adept at mediating between local, intimate rhythms and the rhythms of social as well as environmental structures. If fiction has come to seem ill-suited for such purposes by eco-critics who favor nonfictional genres of writing, it is only because we underestimate what fictional narrative is capable of doing in and with time. As a case study, I consider the work of Virginia Woolf, perhaps one of the most rhythm-conscious of English writers. While her work predates what is generally considered to be the age of globalization, it speaks acutely to concerns that persist today and thus arguably challenges any facile periodizations of globalization, a phenomenon that derives much of its compulsive force from a perpetual sense of surprise at its own novelty. Inventively mediating between biophysical rhythms and social patterns of learned rhythmic repetition, Woolf's work is attuned to the lived experience of the everyday in all its heterogeneity and complexity, its organic and inorganic manifestations, its irreducibly local as well as global patterns. Perhaps most important in the context of my argument is that in its representation of rhythmic variation it offers a radical critique of the only apparent stabilities of the social order and the object world, both of which pretend to a mono-rhythmic harmony that is in fact shot through with discordant and uneasily harmonized rhythms that signify and indeed are themselves the very foundation for potentiality and transformation.

Not dead yet: homogeneous, empty time and the colonized human

Conceptions of time and temporal diversity have tended, in the humanities, to rely upon untenable binaries between an ostensibly homogeneous time of global capital and a heterogeneous temporal diversity located in the margins of modernity or beyond the limits of the human. The theoretical persistence of the concept of "homogeneous, empty time" in writings about temporality and nationalism (Kelly and Kaplan, 2001, pp. 33–42; Lim, 2009, pp. 70–79) has blunted more sophisticated articulations of time in globalization theory, provoking a retrenchment behind irreducible walls of culturally or linguistically defined temporal heterogeneities. Yet the term, "homogeneous, empty time," originated by Walter Benjamin and influentially misappropriated by Benedict Anderson, fundamentally mischaracterizes the nature of time under global capitalism. Disentangling that theoretical misreading is a crucial preliminary step in any attempt to offer a counter-theory of time and globalization.

In his "Theses on the Philosophy of History," Walter Benjamin had argued that "homogeneous, empty time" was a false construct enabling historians to abdicate their responsibility to reclaim and reinvigorate moments of the past that have been lost (Benjamin, 1968). It was, for Benjamin, a rhetorical construct demolished in the very gesture used to invoke it. That act of dialectical negation, lost in Benedict Anderson's borrowing of "homogeneous, empty time" in his influential book, *Imagined Communities* (Anderson, 1983), is captured in Keya Ganguly's discussion of Benjamin's concept which, she writes, is "deployed" only "to augur its negation" (Ganguly, 2004, p. 177). Homogenous, empty time is not and cannot ever be Greenwich Mean Time or Coordinated Universal Time, which are sociological realities that structurally support the movement of global capital yet have never achieved in lived practice the kind of empty ideological hollowness that Benjamin attributed to the rhetorical fiction of homogenous, empty time. The latter is a purely fictitious utopian construct that simply cannot exist rhythmically, which is to say in the lived social practices of sensible and corporeal human beings (Chatterjee, 1999). Insisting on the illegitimacy of homogeneous, empty time as a meaningful lens through which to see the time(s) of the global is an important preliminary move in attempting to rethink how and why time matters in globalization studies. Specifically, it demands that we think more carefully about how and in what locations temporal otherness can be assigned.

Certainly in postcolonial studies the notion that Western or Eurocentric power depends upon and disseminates homogenous, empty time has led to a counter-narrative that attributes an alternative temporal heterogeneity and fullness only to colonized, marginalized, hybridized, or otherwise unassimilated cultural locations. In the work of Homi Bhabha, for example, temporal heterogeneity is less intrinsic to modernity and more the unique performative experience of minoritarian culture (Bhabha, 1994). Ganguly, in her critique of Bhabha's argument, asserts that if "all signs are non-synchronous" one logically has to accept that temporality is heterogeneous for everyone, not just for those ethnic "signifiers" that "one wishes to privilege on account of their postcolonial or subaltern provenance." In the process of shifting the location of temporal "otherness" to the cultural margins and exteriors of modernity, Ganguly argues, "the very notion of non-synchronousness as contradiction within modernity gets lost" (Ganguly, 2004, p. 173). To insist on temporality being irreducibly and incommensurately imbricated with culture, ethnicity and unique location, Ganguly notes, is to ignore Johannes Fabian's warnings in *Time and the Other* that to do so is not to resist but rather to buy wholesale into the European anthropological imaginary which insists that peoples occupying distant geographical spaces somehow occupy unique and non-synchronous times (Fabian, 1983). Recognizing that we are all in the same time frame on the same planet was, for Fabian, the first step toward dismantling racist, evolutionist schemes of manipulation and dehumanization. Yet Fabian's attempt to force his colleagues to recognize their co-evolution and co-existence with every other place and culture smacks, in the contemporary theoretical climate, of a suspect universalism.

Globalization discourse in the humanities tends to be distinguished by a morbid conviction that the homogenous, empty time of capital has so effectively colonized human thought, action, and corporeality that one can only seek alterity beyond the human altogether. There is an eschatological strain in much leftist globalization theory that sees the present as so thoroughly colonized by the time of capital as to leave conceptual space only for a revolutionary time of living so radical as to be incapable of figuration by means of anything but representational caesuras. Eric Cazdyn, for example, is anything but sanguine about the possibility of rethinking or reliving a time outside what he calls the "new chronic." Capitalism's flattened, perpetually extended "meantime" of crisis management without hope of revolution, resolution, or finality is, Cazdyn writes, an "encompassing temporality affecting every aspect of our lives" (Cazdyn, 2012, p. 44). Its global patterns are ultimately indistinguishable from the most intimately local. Its "forms of exchange, meaning-making, social relations, desire, communication, and thought," he writes, "insinuate themselves into our very beings, so much so that attempting to avoid them is like trying to avoid our deepest habits, from the way we hold our bodies to the way we think about how we hold our bodies" (Cazdyn, 2012, p. 196).

Cazdyn's image of resistance, given the totalizing nature of capitalism's penetration into bodily rhythms, is the figure of the undead, atemporal zombie that has enjoyed such a massive resurgence in popular culture. While he calls for a reengagement with the concept of "totality," invoking Masao Miyoshi's "almost unthinkable solution" of "a uniquely transdisciplinary and global environmental protection studies" (Cazdyn, 2012, p. 74), Cazdyn's final figure of the "already dead" zombie emblematizes the hyperbolic endgame of so much writing about globalization in which the hypothetical subject is situated so claustrophobically within a rigidly defined hierarchy of global capitalism that he or she can only evade its nets by disavowing the human altogether. Yet any global system is as dependent upon the rhythmic cycles of local and even microscopic levels as it is upon the slower macroscopic cycles in which it is itself tenuously situated. The rhythms of the everyday are never entirely colonizable. We are not dead yet.

Borrowing from Elizabeth DeLoughrey, who argues for a turn to the sun and the solar as a crucial trope in postcolonial ecocriticism ("like heliotropes," she writes, "we turn to the sun"; DeLoughrey, 2011, p. 235), I would argue for a conception of time in globalization not as a theoretical anachronism colonized by capital or as a principle of intrinsically incommensurable otherness but rather, in its irreducibly rhythmic form, as a governing ecological force that necessarily implicates our local experience with larger global forces of creation, destruction, and resilience. Everyday life is not distant from the global but is rather always and everywhere shot through with its rhythms. This is not to suggest that a conception of a shared global environment and a return to the planetary and cosmological in our formulation of temporality should absolve us from recognizing the very different lived experiences of that globality. What the "international turn" in ecocriticism provokes, though, is attentiveness to the local and to alterity in the context of a shared globality, as Ursula K. Heise has observed in her recent overview of ecocriticism for *PMLA* (Heise, 2013). Another way of thinking through the relations and tensions between local and cosmological time is through Lefebvre's theory of rhythm-analysis.

Lefebvre and rhythm-analysis

Theorists of time from St Augustine onward have yoked temporality to human perception and consciousness in such a way that time has become an unavoidably human category. In the most extreme variant of this idea, time is said to have no existence at all outside a perceiving human consciousness (Barbour, 2001). A long tradition in Western thinking about time has separated human cognition from processes of astronomical movement, geographical formation or autonomous biorhythmic systems. Yet clearly everyday life (a construct with a complicated sociopolitical and theoretical heritage; De Certeau, 1984; Lefebvre, 1991) puts humans in regular and intimate rhythmic relationship with that which is non-human. It would be equally as impossible to live an everyday temporality without the non-human forces of planetary rotation, heliocentric orbit and ecological diversity as it would be to conceptualize temporality without human cognitive processes.

The most compelling, though largely unappreciated, attempt by a major theorist to move beyond purely cognitive models of everyday temporality is the posthumously published work of Henri Lefebvre, *Rhythmanalysis: Space, Time, and Everyday Life* (2004). In this book, Lefebvre conceptualizes the everyday as a theater of multiple and sometimes antagonistic rhythms all lived, tasted and touched in the sensible and corporeal body. To be not only modern but human, Lefebvre suggests, requires living a chaotic and diverse cacophony of temporal rhythms which can be harmonized but are as often prone to breaking out of their confines and producing temporal dissonances, disjunctions and arrhythmias. There is never one time in any one place but rather many times uneasily synchronized. We balance in ways that are often unconscious to us the dictates not of *the* clock (abstractly conceived) but of many discrete clocks in their many locations with the biophysical rhythms both within our own sub-molecular and genetic structures and those of the people around us to whose rhythmic patterns of speech and motion we attenuate our own patterns.

For Lefebvre, human temporality is always and everywhere constrained by biological and cosmological rhythms. "Everyday life," he writes, "remains shot through and traversed by great cosmic and vital rhythms: day and night, the months and the seasons, and still more precisely biological rhythms." In "perpetual interaction" with one another, the "great indestructible rhythms and the processes imposed by the socio-economic organisation of production, consumption, circulation and habitat" are played out and always at stake in the theater of the everyday (Lefebvre, 2004, p. 73). If capitalism constructs its spatial and temporal logic from a fundamental "disdain for the body" and the "time of the living" (Lefebvre, 2004, pp. 51–52), Lefebvre suggests that an attentiveness to the body's polyrhythmic interaction with unique spaces offers a riposte to philosophical and political systems that cannot be "lived, tested, and touched in the sensible and the corporeal" (Lefebvre, 2004, p. 45).

Lefebvre insists that rather than see objects as stable, fixed, and unchanging, we should try to recognize the object world "symphonically" (Lefebvre, 2004, p. 80). While even a flower garden can, to the merely visual sense, present an illusion of "permanence" and "spatial simultaneity" to the point of "sculptural immobility" behind which its rhythms disappear, Lefebvre insists that we "[g]o deeper … listen attentively instead of simply looking … you thus perceive that each plant, each tree, has its rhythm, made up of several" (Lefebvre, 2004, p. 31). "An apparently immobile object," he writes, such as a forest, "moves in multiple ways: the combined movements of the soil, the earth, the sun. Or the movements of the molecules and atoms that compose it." To "the attentive ear," he continues, the object "makes a noise like a seashell" (Lefebvre, 2004, p. 20). Lefebvre thus suggests that time conceived as rhythm avoids the subject-object dichotomy so entrenched in the Western philosophical tradition. Since the body is a "constant reference" for rhythmic time, serving as its "metronome," it governs even the unconscious which, as Lefebvre argues, should be understood as "that which goes on in the body" (Lefebvre, 2004, p. 44). This corporeal temporal assemblage of diverse rhythms necessarily interacts with an object world similarly composed of its own assemblage of rhythms. The divide between a consciousness and its exterior, between the self and its other, becomes blurred by understanding time as a rhythmic orchestration and a configuration of systems.

The great transformative potential in a rhythm is that it never absolutely repeats. "Absolute repetition is only a fiction of logical and mathematical thought," Lefebvre writes (Lefebvre, 2004, p. 7). Repetition, even in mathematics, gives birth to difference, indeed produces it. "There is always something new and unforeseen that introduces itself into the repetitive" (Lefebvre, 2004, p. 6). A critique of the "sculptural immobility" of things and of the social relations that produce them is intrinsic to rhythm. Sameness is always interrupted rhythmically by newness, even in (or perhaps especially in) the recurrence of cosmic or biological patterns. “Dawn always has a miraculous charm,” Lefebvre writes, “hunger and thirst renew themselves marvellously” (Lefebvre, 2004, p. 73). Each day introduces some variation on repetition which is how newness enters the world. Lefebvre’s ecstatic and quasi-metaphysical sense of the profound transformative potential latent within the immobility of the object world reflects his intellectual and political heritage within the Western Marxist tradition, according to which dialectical thought (a term oversimplified and reviled in the current trend of anti-Hegelianism in the humanities) exposed a radical instability within not only the current social order but also the object world itself. For Herbert Marcuse, for example, reality was made up of “fatal contradictions” that dialectical thinking needed to expose in order to reveal the “promises and potentialities” of life under the dictates of established “common sense” (Marcuse, 2007, p. 64). A struggle with the “conditions of … existence,” manifested consciously in humanity, is blindly and unconsciously present in “inorganic matter” for Marcuse as well (Marcuse, 2007, p. 65). Lefebvre translates this dialectical sense of the transformative potential of reality into temporal terms, insisting that temporal rhythms obscured in the ersatz harmonies of modernity nevertheless remain powerfully fraught with discordance and dissonance just underneath the surface of reality.

Panarchy and rhythm

Lefebvre’s rhythmic understanding of the relationship between the local and the global has no analogue within the humanities or social sciences, and despite the influence on cultural studies of his landmark *Critique of Everyday Life* or the impact of his *The Production of Space* on Marxist geographers like David Harvey and Neil Smith, Lefebvre’s *Rhythmanalysis* has attracted little attention.[1] Yet its unique appreciation of the potentialities inherent in the rhythmic character of everyday temporality makes it a provocative theoretical and political counterpart to the studies of rhythm and transformation that characterize ecological resilience studies in the natural sciences. If Lefebvre’s late writings on rhythm (composed when the thinker was in his seventies and eighties) can appear unpalatably mystical and quasi-poetic, their focus on the intimate and intricate relationships between local and global rhythmic patterns are equally as characteristic of the “hard science” carried out empirically and theoretically by scholars who study resilience, sustainability and adaptability in ecological and social systems under the banner of the Resilience Alliance.

Resilience studies, generally considered to have been inaugurated by the Canadian ecologist C.S. Holling in the 1970s, recognizes as a core theoretical tenet the necessarily interdependent relationships among natural processes that occur at radically different spatial and temporal scales. These “cross-scale interactions,” which range “from centimeters and days in the dynamics of photosynthesis through kilometers and decades for disturbance processes that shape patterns on landscapes to hundreds of kilometers and millennia for geomorphological processes," occur in human affairs as well, "from the individual to the community, to the nation and region, and to international patterns of relationships" (Gunderson and Holling, 2002, p. xxii). Understanding the threat to the resilience of natural systems demands a world view that refuses to decouple microcosmic temporal patterns from macrocosmic ones, or natural from social systems.

While ecological processes can obviously be understood spatially, Holling's innovation was to propose a model of resilience and adaptation that is fundamentally temporal in its conception. Its graphic representation is that of the infinity symbol, the loops of which suggest not the eternal changelessness of natural systems but rather the ceaseless processes of transformation, destruction, reorganization, and retention that are integral to the adaptive survival and vitality of those systems. More properly speaking, resilience theory's model is not that of a single infinite loop, but rather that of a series of infinite loops at various levels, with the small and fast rhythms of change nested within a series of increasingly larger and slower scales of adaptive change. This model, importantly and evocatively for resilience theory, is not understood to be describing a hierarchy, a term "burdened by the rigid top-down nature of its common meaning," but rather a "panarchy," a neologism drawing on the figure of the Greek god Pan who embodies both a creative as well as a "destabilizing, creatively destructive role" (Gunderson and Holling, 2002, p. 74). The smaller local rhythms of agents and processes within the panarchy are not controlled by or simply at the whim of larger global rhythms. Rather the relationship between the smaller and larger, the faster and slower, is mutually constitutive and enabling. Generally, "larger and slower components ... provide the memory of the past and of the distant to allow recovery of smaller and faster adaptive cycles" (Gunderson and Holling, 2002, p. 20), while "attributes of the slower levels emerge from experience of the faster" (Gunderson and Holling, 2002, p. 72).

At the forefront of ecological research, then, is a conception of the relationship between the local and the global that is inescapably rhythmic. As for Lefebvre, the challenge provoked by a recognition of this relationship is not a passive one in which we are invited to learn to listen symphonically to everyday rhythms or simply to appreciate the rhythms of resilience, adaptation and rebuilding that happen over large temporal scales in an ecosystem. The point of rhythm-analysis, to borrow Marx's famous 11th thesis on Feuerbach, is not only to hear rhythms but to change them as well. Resilience studies could very well be politically quiescent, communicating the message that any ostensible environmental calamity is simply one loop on the infinity symbol of the panarchy, providing the larger ecosystem with the potential to better adapt and restructure itself for the next cycle. Yet this is not the case. The Resilience Alliance recognizes that human rhythms have been destructive beyond an ecosystem's ability to adapt. The thrust of their interdisciplinary work is to better understand how panarchies adapt and survive in order that humans might better organize their own rhythmic involvement in the panarchy in such a way as to encourage and foster resilience rather than collapse. Underlying every insight it provides into an ecosystem's capacity for resilience is the knowledge that humans can and have eroded that capacity and that "the capacities for self-repair of ecosystems can no longer be taken for granted" (Folke et al., 2004, p. 575).

At the same time, the capacity for humans to carry out transformative and radical change in the rhythms of everyday life and in the various panarchies in which that everyday is involved becomes exponentially greater when we model globalization as an assemblage of intricately inter-involved rhythmic patterns as did Lefebvre or as does resilience studies. In this way, we begin to imagine ways in which we can resist the grim prognosis for human action and capacity that follows from conceptual models of globalization which, in relying upon theoretical oversimplifications of time and representing global capital as exerting homogenous and tentacular control over every aspect of temporality, thereby relegate humanity to a moribund state of recognition of its own putrefaction and zombification.

Narrative and the rhythms of the global

The struggle of globalization, as Eric Cazdyn notes, is "a struggle over how to tell the story of globalization" (Cazdyn, 2012, p. 114). Narrative is able to mediate the temporal divides between larger global cycles and smaller idiosyncratic local rhythms. Yet fictional narrative has come to seem ill-suited to such a project. It is this recognition of the limitations of fiction and literary narrative, which inevitably "privilege the human voice and human agency," that has traditionally prompted eco-criticism to favor nonfictional genres in its analysis (Heise, 2013, p. 641). Yet to understand fiction as invested solely in the representation of limited human consciousness and agency is to miss the presence of rhythm in narrative. Integral to fictional narration is a representation of human rhythms, which as Lefebvre writes, "insert us into a vast and infinitely complex world" (Lefebvre, 2004, p. 82). The relationship between "our" rhythms and those of the "complex world" around us is central to Lefebvre's understanding of a temporality that is not restricted to the subjective or to the anthropocentric but is rather a privileged mediator, if we know how to "listen," between the human and its others, between the local and the global.

As Lefebvre's work suggests, rhythmic time necessarily breaks down the barriers between subjectivity and the object world or between individuals and larger collectives or systems. In her study of time in fiction, Elizabeth Deeds Ermarth identifies this feature of what she calls "postmodern" temporality, but which I would argue is by no means restricted to a particular period or genre of narrative. Rhythmic time, as constructed by novelists like Julio Cortázar, performs an "incantatory alteration between subjects, systems, sets of consistencies," privileging not the "transcendent existence of the one who rationalizes" but rather a "conjugation" of rhythmic groups (Ermarth, 1992, p. 47). The "individual consciousness" is replaced by a more "multivocal and systemic identity" defined by its rhythmic interactions (Ermarth, 1992, p. 53). Equating this rhythmic postmodern temporality with the improvisatory rhythms of jazz, Ermarth sees in Cortázar's riff on the Cartesian premise of existence ("I swing, therefore I am," Cortázar writes) the promise of a subjectivity defined through its rhythmic interactions rather than through its claim to consciousness.

If we do not know how to listen to a text that narrates these rhythmic mediations, we might be inclined to categorize that text as being outside the boundaries of narrative itself, as does Paul Ricoeur with Virginia Woolf's *The Waves* in his *Temps et Recit* (Ricoeur, 1985, pp. 97–98). Ricoeur's three-volume critical opus on time and narrative does much to obscure fictional narrative's rhythmic imperatives, instead situating the temporality of narrative squarely in the tradition of Western philosophical conundrums about time and human perception dating back to Augustine and Aristotle. Narrative for Ricoeur enables and makes productive "thought about time," serving as human time's "guardian" in that "there can be no thought about time without narrated time" (Ricoeur, 1988, p. 241). By narrative, however, Ricoeur means very specific kinds of "tales about time" involving characters, plots, and subjective reflections that ultimately reinforce chronology and human cognition. While he uses Woolf's *Mrs. Dalloway* as an example of narrated temporality, he throws up his hands (not unlike many readers) at the temporality that her more experimental text *The Waves* enunciates. The best he can do is exclude this "polyphonic novel" from consideration altogether. *The Waves* is "no longer a novel at all" but a "new genre," he writes (Ricoeur, 1985, pp. 97–98). It presents consciousness in a novel way but is excluded from the genre of narrative because in crossing the generic threshold that it does, it no longer serves as a "tale about time." Yet certainly *The Waves* is nothing if not a narrative about time. Not only is it about time, but its structure and style make temporal rhythms central to its project of situating the human in relation to larger non-human forces and contexts. Nor is it unique in relation to other narratives in that sense. Rather its idiosyncratic style exaggerates a central component of narrative that we might be more inclined to observe were we to stop thinking of fictions as philosophical puzzles for working out Augustinian temporal paradoxes and begin reading them for the ways in which they mediate the complex array of temporal rhythms that constitute everyday life.

In order to understand how fictional narrative works to mediate the rhythms of the local in relation to the global, or in other words to negotiate the panarchy within with the human is situated, we need to appreciate how even a text known for its emphasis on individual subjective human perception or the so-called "stream of consciousness" technique necessarily and integrally represents human rhythms as part of a larger whole outside the subjective or cognitive. It is for this reason that I will focus my textual analysis here not only on *The Waves* but also on Woolf's earlier novel *Mrs. Dalloway*, a text often understood to exemplify early 20th-century modernist literature's focus on individual consciousness and subjectivity.

In some ways, it is difficult to imagine a text less suited on the surface for a discussion of narrative's relation to globalization. Written and published before World War II, set entirely in a London still bristling with nationalist and imperial sentiment and with a plot revolving around a politician's wife's preparations for an elaborate dinner party, the text seems quaintly insulated from the pressures of global capitalism. Yet, as I have argued elsewhere, the instantiation of World Standard Time in the period during which *Mrs. Dalloway* was written and with which the novel is insistently preoccupied, heralded and set in place the foundations for globalization as we now understand it (Barrows, 2011). This seemingly parochial novel of pre-globalization, in its stylistic insistence upon the uneasy rhythmic relations between local and global patterns of experience, in fact illustrates narrative's power to intervene in the panarchy at a time when the structural foundations were being laid for the rise of globalization and its claims of mono-rhythmic global harmony.

"Time flaps on the mast": Virginia Woolf and the rhythms of the everyday

Virginia Woolf is arguably one of the most rhythm-conscious prose writers in the English language but her reputation as a writer obsessed with subjective, interior, mental processes seems to make her an unlikely choice for a study of narrative and the global. As Melba Cuddy-Keane argues in her discussion of "global consciousness" in Woolf's writing, while Woolf may "not write extensively and explicitly about travel, geography and cross-cultural encounters" (Cuddy-Keane, 2010, p. 158), she does evoke "through landscape, wind or sky" a "global consciousness" that functions as a "backdrop against which human figures fluctuate and change," and that situates her as a writer "in a greater and vaster world" and as "part of the larger whole" (Cuddy-Keane, 2010, p. 174).

I would agree with everything but Cuddy-Keane's choice of the word "backdrop," which places the global in the backstage of Woolf's work as simply a cloth against which the properly foregrounded humans stand in high relief. Rather than understanding Woolf's humans as situated in front of the global, a rhythmic reading of her work conceives them, in Lefebvre's terms, as intrinsically implicated in the global by means of their rhythmic interactions. In books like *Three Guineas*, Woolf wrestled quite rigorously with what she took to be the great ethical and aesthetic problem of finding a means to mediate the seeming divide between global crises like the war and the everyday lives of women. The former, which is figured repeatedly in *Three Guineas* by photographic images of corpses and bombed homes ("dead bodies and ruined houses"), is bound up, as Woolf endeavors to show throughout the text, with the latter (Woolf, 1968, p. 154). Woolf's thesis and rallying cry is that "the public and the private are inseparably connected" (Woolf, 1968, p. 258). The public spectacles of photographically captured "dead bodies and ruined houses," ostensibly so disconnected from the demands of private life, in fact demonstrate to Woolf the unity of "the public and the private, the material and the spiritual." Her vision in the conclusion of *Three Guineas* is of "one world, one life" and of a human spirit that has the capacity "to overflow boundaries and make unity out of multiplicity" (Woolf, 1968, p. 259).

In her narrative fiction, that vision of unity and capacity of the spirit is best figured by means of rhythm. In a letter to Vita Sackville-West, Woolf explained the central importance of rhythm in her compositional style. "Style is a very simple matter," she wrote, "it is all rhythm ... Now this is very profound, what rhythm is, and goes far deeper than words. A sight, an emotion, creates this wave in the mind, long before it makes words to fit it; and in writing (such is my present belief) one has to recapture this, and set this working" (Sutton, 2010, p. 176). Rhythm for Woolf, as Emma Sutton argues, defies definition in part because of its non- or pre-linguistic nature. "Woolf variously imagines rhythm as natural or corporeal, aesthetic and historically specific, or both," writes Sutton (2010, p. 178). In her "Letter to a Young Poet," Woolf had described the "instinct of rhythm" as "the most profound and primitive of instincts" (Stonebridge, 1995, p. 99). In its primitivism, it is both dangerous and potentially liberating for subjectivity, as Lyndsey Stonebridge argues. In *The Waves*, a novel that notoriously frustrates readers by its presentation of multiple subjective monologues from six different characters, the focus is arguably not on subjectivity or on the individual characters at all. As Susan E. Lorsch explains, Woolf in the novel "pulls the six little fish out of the cauldron in order to work back to the level of generalization, to speak to us of that 'globe of life' itself ... characterization and story in *The Waves* are the means to another end: rhythm" (Lorsch, 1979, p. 201).

The Waves exemplifies Woolf's desire to relate the intimate biorhythms of corporeal life to larger global rhythms, whether those of the world of international commerce in which the character Louis works, or those of the planetary cycles that produces the seasons, night and day, and the action of the tides. The tension established in the text between imperial images of global connectivity and images of ecological rootedness in soil, stalk, flower, and sea foam is quite deliberately sustained. Just as her characters insist upon a reified vision of Percival (the character who dies as a servant of empire in India and who remains voiceless throughout the novel) as a force of civility amidst the chaotic forces of the larger globe, Woolf in the interlude passages equates the action of the waves with stereotypical images of a violent Orientalism, as when in the third interlude the waves drum on the shore "like turbaned warriors" with "poisoned assegais" (Woolf, 1931, p. 75). Lacking a language or imagery to capture a rhythmic interrelatedness beyond "the common mainspring" of clockwork synchronization (Woolf, 1931, p. 94), Woolf underscores the human tendency to fall back upon stale and arid racial or political categories as placeholders for a rhythmic interrelatedness that she knows to be far more unstable and ephemeral. *The Waves* ends on a morbid note of false triumph, with the desires of its characters to find meaningful rhythmical relationships within the panarchy translated into Bernard's quixotic and overblown closing apostrophe to Death. Existing in a strange otherworldly register, as if it were a symphony of disembodied voices reflecting spectrally on the significance of life rather than being an actual representation of life, *The Waves* expresses more powerfully than any other of Woolf's novels the urgency of rhythmically relating the local and the global, without itself ever finding a convincing means of formally mediating between the two.

An earlier novel like *Mrs. Dalloway*, though less explicitly and hyperbolically obsessed at the thematic level with the rhythmical relationships between micro and macro processes and with the human relationship to the nonhuman world, arguably demonstrates more powerfully and characteristically the possibilities of narrative in articulating the polyrhythmic densities of the everyday and to explore the possibilities of interrupting and exploding rhythmic hierarchies. Multiform and discordant rhythms of everyday life echo throughout *Mrs. Dalloway*. Woolf's rendering of a day in June captures not only steady rhythmic patterns in the life of London but also the arrhythmic collapse and rebuilding of those regularities. The linkage between unique biorhythms such as Clarissa Dalloway's irregular heartbeat, mentioned in the first pages of the novel, and the rhythmic organization of urban space, usually invisible for the perfectly adjusted organism, becomes transparent for the ill or the maladjusted body. This connection between the patterned movements of individuals and the greater sonic patterns of urban space becomes momentarily apparent to Clarissa when she observes the woman in the flat across from her, who seems to move in coordination with the bells of Big Ben "as if she was attached to that sound, that string. Gigantic as it was, it had something to do with her" (Woolf, 1925, p. 127).

The character Peter Walsh earlier in the text similarly finds himself coordinating his footsteps with the sound of a band of boys in uniform marching up Whitehall. He hears the approach of the regimented boys as a "rustling, regular thudding sound, which as it overtook him drummed his thoughts, strict in step, up Whitehall, without his doing" (Woolf, 1925, p. 51). The fact that Peter so quickly loses volition in regards to his thoughts and his stride indicates how hypnotic and compelling is the temptation to lose oneself in the lock-step of powerfully coordinated rhythms. Despite the seductive compulsion of such rhythmic uniformity, the human body does not easily conform to rational or mechanical imposition. Unless drugged and beaten by habit, its tendency is stridently to resist uniformity, compliance, and synchronization. Even Walsh, so predisposed to admire discipline and training, simply cannot keep his aged body up to pace with the boys. "I can't keep up with them," he thinks, and indeed Woolf seems to suggest that nobody (no body) can keep pace with this discipline (Woolf, 1925, p. 51).

"Rhythm appears as regulated time, governed by rational laws," Lefebvre writes in *Rhythmanalysis*, "but in contact with what is least rational in human being: the lived, the carnal, the body" (Lefebvre, 2004, p. 9). There is, then, something inherently resistant to external rhythmic discipline in the body itself, which may explain why Doctors Holmes and Bradshaw in Woolf's novel attempt to manage the shell-shocked war veteran Septimus Smith's dangerous arrhythmic condition through tactics of bodily training and control. Segregating Septimus and disciplining his errant body is a means of enforcing, through rhythmic intervention, a false harmony or eurythmia on that body. Left to his own devices, however, Septimus experiences both the beauty and the horror of a world unleashed from its ersatz harmony and revealed in the diversity of its interrelated but never quite harmonious rhythms: the world of the early 20th century as one might hear it in avant-garde, experimental, or atonal orchestral music. The "new religion" that Septimus dimly perceives as a sonic phenomenon is one that promises to dismantle the kinds of rigid "sculptural" immobilities that accrete through the repetition of habitual patterned behaviors. As Lefevbre reminds us, the great potential in rhythm is that it never absolutely repeats. While this is often frightening to the pathological Septimus who is surrounded by nightmarish transmogrifications of the familiar (an old woman's head in a fern, a dog becoming a man), the novel itself endorses and replicates a vision of the world wherein objects and social relationships are always rhythmically unstable and fluid.

One of Woolf's many recurring metaphors for the changeability and potentiality intrinsic to daily rhythms is in the action of the clouds, which seem superficially stable but are in fact shifting and ephemeral. In the section of the novel where Elizabeth Dalloway is wandering alone down The Strand after her tea with Miss Kilman, the narrative pauses to consider the effect of the clouds momentarily covering the sun:

> although the clouds were of mountainous white so that one could fancy hacking hard chips off with a hatchet, with broad golden slopes, lawns of celestial pleasure gardens, on their flanks, and had all the appearance of settled habitations assembled for the conference of gods above the world, there was a perpetual movement among them. Signs were interchanged, when, as if to fulfil some scheme arranged already, now a whole block of pyramidal size which had kept its station inalterably advanced into the midst or gravely led the procession to fresh anchorage. Fixed though they seemed at their posts, at rest in perfect unanimity, nothing could be fresher, freer, more sensitive superficially than the snow-white or gold-kindled surface; to change, to go, to dismantle the solemn assemblage was immediately possible; and in spite of the grave fixity, the accumulated robustness and solidity, now they struck light to earth, now darkness.
>
> (Woolf, 1925, pp. 138–139)

The only apparent fixity, gravity, and solemnity of the object world dissolve at a mere "puff of wind" (Woolf, 1925, p. 138). The shifting cloud formations signify the tenuousness of other more artificial and self-enforced types of rhythmic disciplining, as earlier in the novel when a moment of silence falling on London and "on the mind" is compared to a cloud crossing the sun: "Time flaps on the mast," Woolf writes. "Rigid, the skeleton of habit alone upholds the human frame" (Woolf, 1925, p. 49). Elsewhere, the city of London itself at evening metaphorically "pinions" the sky, imposing its own newly created daylight savings time ("Mr. Willett's summer time") of "prolonged evening" onto the natural waning of light and heat and color in the sky: "I resign, the evening seemed to say, as it paled and faded above the battlements and prominences … but London would have none of it, and rushed her bayonets into the sky, pinioned her, constrained her to partnership in her revelry" (Woolf, 1925, p. 161).

What Septimus's morbidly pathological condition reveals to him in a heightened fashion is what Woolf or Lefebvre would argue is intrinsic to everyday life: its inherent potential for transformation, renewal and reform, its uneasy compromise between incompatible rhythmic patterns. In her recent book, *In the Meantime: Temporality and Cultural Politics*, Sarah Sharma characterizes the temporality of everyday life as "differential, relational, and tangled" (Sharma, 2014, p. 80). Resisting the rhetoric of a globalization discourse that insists upon a universally experienced temporal acceleration, Sharma investigates the tangled and multiple power relationships among groups of people within the same temporal grid whose rhythms are quite variable and distinct depending on their relationship to established structures of power and privilege. Narratives such as *Mrs. Dalloway* reveal fiction's investment in inhabiting, timing, and exploring the fault lines in those differential and tangled temporal relationships. This investment is by no means unique to "postmodern" or formally experimental texts. Any narrative with a legitimate claim to verisimilitude must attempt some degree of mapping the differential and competing rhythms that constitute everyday life. In the fiction of Virginia Woolf, these tangled, competing rhythms are made integral to the compositional style, the metaphorical imagery, the thematic content, indeed to the entire narrative enterprise. "It is all," as she wrote, "rhythm."

Understanding temporality as rhythmic in Lefebvre's or in Woolf's terms allows us to apprehend the rhythmic alterity that is fundamental to lived temporal experience. If Lefebvre makes clear why it is important to understand the politics of temporality through rhythm, Woolf's body of work demonstrates how such a temporality could be spoken (or could "swing" in Cortázar's jazz lexicon). The subject that emerges at the intersection of globalization discourse and transnational ecocriticism will be recognizable by its rhythmic temporality: a time that in its heterogeneity and multiform diversity can never adequately be captured by purely conceptual constructs like "homogenous, empty time" nor ever entirely colonized by the time of global capital.

While Barbara Adam warned that the "industrial timescape" leaves "no temporal feature untouched" in its regulation and disciplining of the "pace and seasonality of organisms and beings" (Karsten, 2013, p. 353), rhythm-analysis reminds us that the "homogenization of time patterns" that accompanies the development of a global economy is not capable of "colonizing the whole globe," as Luchien Karsten suggests (Karsten, 2013, p. 325). Yet Woolf's own meditations on rhythm as a primordial precognitive force "beyond words" gesture to the difficulties inherent in a rhythm-analytical praxis. Rhythms affect us often without our knowledge or volition and could even be understood as merely symptomatic traces of structural or institutional forces that are beyond the reach of the rhythm-analytical. Bringing the rhythmic relationships of ecological panarchies into the realm of consciousness can by no means be confused as a form of political practice in itself. Yet at the same time no useful social, political, or ecological praxis can be derived from conceptions of time that fundamentally misrepresent the rhythmic linkages and interdependencies between the human and the global. Rhythm-analysis, despite these constraints, reveals latent and volatile potentialities within the times of the local and the everyday to shape and redirect the rhythms of the global.

Note

1 One exception to this general disregard is in the field of cultural or human geography. Two relatively recent collections of essays, one entirely devoted to Lefebvre, have explored the implications of rhythm-analysis on geography (see Edensor, 2010; and Mels, 2004).

References

Anderson, B. (1983) *Imagined Communities.* London: Verso.

Barbour, J. (2001) *The End of Time: The Next Revolution in Physics.* Oxford: Oxford University Press.

Barrows, A. (2011) *The Cosmic Time of Empire: Modern Britain and World Literature.* Berkeley, CA: University of California Press.

Benjamin, W. (1968) *Illuminations.* New York: Schocken Books.

Bhabha, H.K. (1994) *The Location of Culture.* London: Routledge.

Cazdyn, E. (2012) *The Already Dead: The New Time of Politics, Culture, and Illness.* Durham, NC: Duke University Press.

Chatterjee, P. (1999) Anderson's Utopia. *Diacritics*, 29(4), pp. 128–134.

Cuddy-Keane, M. (2010) World Modelling: Paradigms of Global Consciousness in and around Virginia Woolf. In Shahiari, L. and Potts, G. (eds) *Virginia Woolf's Bloomsbury Volume 2: International Influence and Politics* (pp. 158–176). New York: Palgrave.

De Certeau, M. (1984) *The Practice of Everyday Life.* Berkeley, CA: University of California Press.

DeLoughrey, E. (2011) Heliotropes: Solar Ecologies and Pacific Radiations. In DeLoughrey, Elizabeth and Handley, George B. (eds) *Postcolonial Ecologies: Literatures of the Environment* (pp. 235–253). Oxford: Oxford University Press.

Edensor, T. (ed.) (2010) *Geographies of Rhythm: Nature, Place, Mobilities and Bodies.* Surrey: Ashgate.

Ermarth, E.D. (1992) *Sequel to History: Postmodernism and the Crisis of Representational Time.* Princeton, NJ: Princeton University Press.

Fabian, J. (1983) *Time and the Other: How Anthropology Makes its Object.* New York: Columbia University Press.

Folke, C. et al. (2004) Regime Shifts, Resilience, and Biodiversity in Ecosystem Management. *Annual Review in Ecology, Evolution, and Systematics*, 35, pp. 557–581.

Ganguly, K. (2004) Temporality and Postcolonial Critique. In Lazarus, Neil (ed.), *The Cambridge Companion to Postcolonial Literary Studies* (pp. 162–179). Cambridge: Cambridge University Press.

Gunderson, L.H. and Holling, C.S. (2002) *Panarchy: Understanding Transformations in Human and Natural Systems.* Washington, DC: Island Press.

Heise, U.K. (2013) Globality, Difference, and the International Turn in Ecocriticism. *PMLA*, 128(3), pp. 636–643.

Karsten, Luchien (2013) *Globalization and Time.* London: Routledge.

Kelly, J.D. and Kaplan, M. (2001) *Represented Communities: Fiji and World Decolonization.* Chicago, IL: University of Chicago Press.

Lefebvre, H. (1991) *Critique of Everyday Life Volume 1.* London: Verso.

Lefebvre, H. (2004) *Rhythmanalysis: Space, Time, and Everyday Life.* London: Continuum.

Lim, B.C. (2009) *Translating Time: Cinema, the Fantastic, and Temporal Critique.* Durham, NC: Duke University Press.

Lorsch, S.E. (1979) Structure and Rhythm in The Waves: The Ebb and Flow of Meaning. *Essays in Literature*, 6, pp. 195–206.

Marcuse, H. (2007) *The Essential Marcuse: Selected Writings of Philosopher and Social Critic Herbert Marcuse.* Boston, MA: Beacon Press.

Mels, T. (ed.) (2004) *Reanimating Places: A Geography of Rhythms.* Surrey: Ashgate.

Ricoeur, P. (1985) *Time and Narrative Vol. II.* Chicago, IL: University of Chicago Press.

Ricoeur, P. (1988) *Time and Narrative Vol. III.* Chicago IL: University of Chicago Press.

Sharma, S. (2014) *In the Meantime: Temporality and Cultural Politics.* Durham, NC: Duke University Press.

Stonebridge, L. (1995) Rhythm: Breaking the Illusion. In Wussow, Helen (ed.), *New Essays on Virginia Woolf* (pp. 99–115). Dallas, TX: Contemporary Research Press.

Sutton, E. (2010) Putting Words on the Backs of Rhythm: Woolf, "Street Music" and The Voyage Out. *Paragraph*, 33(2), pp. 176–196.

Woolf, V. (1925) *Mrs. Dalloway.* New York: Harcourt.

Woolf, V. (1931) *The Waves.* New York: Harcourt.

Woolf, V. (1968) *Three Guineas.* London: The Hogarth Press.

11 Checked baggage

An afterward

Sarah Sharma

There is a difference between the traveler who must check their baggage and the one who carries it on. The latter is always assumed to be traveling light, everything on their person ready at hand. These days traveling light is synonymous with traveling smart, but there is no real way to travel light when you are writing about time, let alone time *and* globalization. Writing about time and globalization is laden and heavy, but also distracting, and with a lot of disciplinary baggage. I know this from firsthand experience.

My own work on time and globalization is concerned with the cultural fixation on the speed of life and how the experience of speeding up is too often imagined as if it were a universal and global human experience. Beyond the popular imaginary, many critical treatments of globalization quite often lament the loss of time or time-orientation as a byproduct of accelerated processes of global capital. I have argued that this overwhelming focus on tempo has obscured the multitude of time-based experiences and the multiple temporalities that actually do compose the social fabric (see Sharma, 2014). Instead of an individualistic and universalizing treatment of time, I argue for a political sense of time understood as always already relational and in this way necessarily collective.

Yet, my ability to conjure and outline further the contours and possibilities of this political sense of time is quite often stymied, impeded by at least two pieces of oversized "baggage." On the one hand, I am usually greeted by questions related to an individual's own personal time management anxieties, their ability to manage, balance, or find more time. On the other, it is quite common to be asked to engage publicly in what others have said about time in every field, from physics to philosophy to philology. When I speak of time in academic settings I am simultaneously greeted with "do you mean Bergson's duration, do you cite Kant, or do you mean Heidegger?" This is then quickly followed by a FitBit-adorned wrist shooting up into the air to ask, "what do you think of productivity apps?" While both sets of questions are great for conversation, they delimit thinking about the cultural politics of time. One line of questioning is completely individualistic and the other somewhat abstracted from everyday life. Neither line of inquiry can account for the differential and inequitable social experience of time.

Every field has its resident time people and a set of key time pieces, and every person has problems managing time. It does run out, after all. To be concerned with globalization brings with it its own set of intellectual and practical barriers. The risk of engaging globalization from new perspectives is palpable. What new and important things can be said about this constantly contested and uneven social reality?

Enter this volume – significantly unencumbered, yet not entirely free from the weight of disciplinary baggage, it accomplishes and introduces an interdisciplinary critical time perspective into globalization studies, wherein time is treated as an analytical lens. In this collection, time emerges very distinctly as a bit of an unruly rupture. It offers a new site of analysis and at the same time a new target for political possibility within the timescapes of uneven globalization processes. It is precisely in response to a world/society proceeding with the baggage of uncertainty that the contributors in this volume must wrestle with what to lug around, what to pack and what to leave behind, from their own respective disciplines. Sometimes the actual weight of living in a precarious world exceeds and challenges the conceptual boundaries that have been imposed when maintaining discrete disciplinary fields, but this volume treats this tension with the utmost care and attention. The volume never demands of its contributors upfront to utilize an agreed upon set of definitions of time. To have done so would be responsible and careful, an earnest attempt at grappling with the true complexity of what is on the table here, but what most likely would have emerged would have been a well-rehearsed re-hashing of a set of debates and arguments, ones that do not promise new conceptual terrain. As a reader, it would have offered as much pleasure and insight as the experience of standing around an airport baggage turnstile staring at everyone else's retrieved bags while waiting for your luggage to come.

The ultimate strength of the volume lies in its ambivalence toward an origin story. Each piece brings forth a temporal perspective that makes strange the normalizing critiques of globalization within the authors' respective fields. Freed of a delimiting definitional origin story for "what is time," time is instead treated as having a rhythmic conceptual and experiential terrain itself that alters how one approaches globalization. What constitutes the rubric of time here is vast and far-reaching. The temporal terms that compose the pages include: the past, the future, the present, tempo, speed, pace, minutes, real time, flying time, night time, the dreaming hours, inertia, slowness, unrealized pasts and futures that will not arrive, bodily time and labor time. These are just a few of the things we talk about when we talk about time.

The temporal terms and conditions scattered through the pages are held together by a rather powerful sentiment. In reading this collection I was struck by how each chapter acknowledges and is resistant to both the state and market discourses of globalization that are exerting control over macro- and micro-political processes of everyday life. Yet, this was not necessarily the entry point of critique. Instead, these authors are also equally tired of the catastrophic discourses of globalization within their respective fields. Such catastrophic discourses are responsible for maintaining a universalizing treatment of time or forgetting the importance of time altogether (chapters by Orpana, Barrows, Birth). As Adam Barrows points out, "no useful social, political, or ecological praxis can be derived from conceptions of time that fundamentally misrepresent the rhythmic linkages and interdependencies between the human and the global." Barrows captures exactly what the volume succeeds in doing. It seeks and then it unravels the rhythmic linkages.

A chapter on Virginia Woolf (Barrows) has urgent implications for a chapter on global pandemics, namely SARS (Zhou and Coleman). A chapter that forefronts the biological costs of 24/7 living (Birth) is both critical of theory's blindness to the biological but then complements a chapter on Yugoslavism (Rethmann). In her chapter, Rethmann locates Yugoslavism as encapsulating the provocative function of hope and argues for the need to cultivate temporal bridges in order to re-activate political activism. If Birth says we forget about time and its costs, Rethmann shows we forget about time and its potential. These two chapters meet compellingly in the realm of sleep deprivation (Birth) and the political need for dreams (Rethmann). Birth warns us that to treat all hours of the day as equivalent is not only biologically but also politically unsound. Zhou and Coleman remind us that the contents of a passenger plane flying at high speeds across the globe are not containable in any uniform way. Spilling out are the multiple temporal values and currencies of monies, people and goods, but also the unquantifiable and invisible byproducts of networks where infections become global pandemics that follow the itinerant paths of global cities. These infections that might just materialize intersect with the politics of pre-emptive security that Stockdale argues in this volume are "conceptually incoherent, since its constitutive temporalities compromise its capacity to deliver 'security' as defined under its own normative framework." For Stockdale this then demands a construction of the present as always inherently insecure.

Reading Zhou and Coleman with Stockdale it is not hard to surmise that against the backdrop of an insecure present, certain temporal fates are more likely to occur then others. Read together one cannot help but think about how the work of protecting already insecure populations against this always already insecure future too often recedes from view. Pre-emptive security privileges some populations and ideologies over others. For Bellamy, in referencing oil futures, he recognizes that the necessary rhythmic variation of catastrophe and crisis also requires a unique temporalized political response by way of slow friction, or more specifically the necessity of entropy. For Zhou and Coleman the possibility of viruses going anywhere in no time reveals the limits of a conception of bounded and sovereign space, something Hassan also questions but in terms of the liberal subject and the tempo of democratic processes. Both Hassan and Zhou/Coleman argue for a temporal strategy to attend to new temporal realities. Hassan suggests, "recognition of the fuller spatio-temporal dimensions of political sovereignty will give a fresh perspective on neoliberal globalization and suggest some ways forward from the political impasse that post-modern capitalism creates for us today." For Zhou and Coleman, their temporal strategy requires anticipatory recognition of unpredictable rhythms of networks and their material effects.

Time is a lens, a target and an unexhausted field of political possibility. Time can be re-seized in new ways by those subjects who have been the exhausted targets of bio-inclusive labor (Orpana). Solidarity must be temporalized, synchronized transnationally, to occur at the very same time irrespective of time zones in order for the global linkages of coeval agency to emerge (Hope). These calls for action and solidarity are attuned to the rhythmic variation of different temporal horizons of political possibility. Some of them recognize that exhaustion is differentially experienced along the lines of gender, class and geography, for example (Orpana and Hope).

I have argued before that time should be treated as an intersectional type of social difference rather than an addendum to singular identity categories (Sharma, 2014). Time is lived at the intersection of a range of social differences that include class, gender, race, immigrations status, and labor. A temporal perspective offers insight into inequality that recognizes the transecting multi-dimensionality of social differences. Treating time as an isolatable characteristic of difference rather than a relation of power can lead to violent ideological renderings of others, for example in the label "colored people's time" or within the developmental discourse that attributes much of the geography of the world subject to "primitive time." Instead, I treat time as a lived experience, always political, produced at the intersection of a range of social differences and institutions, and of which the clock is only one chronometer.

Relationships of domination are exercised in time and by time. In order for this to become visible, the study of time must be politicized with an eye toward temporal difference. This necessarily requires an interdisciplinary perspective. The current and persistent contradictions of uneven globalization demand rhythmic variation itself as a political project. Perhaps Critical Temporality is a new discipline on the horizon. Perhaps this project is emblematic of such a turn. Its most minor contribution is grand enough: it is a corrective to the dominance of space within critical approaches to globalization. Moreover, it is a corrective to the focus on speed and time-space compression as the only ways of thinking about time and globalization. However, its grandest gesture is enlivening rhythmic variation across everyday life and across the disciplines.

In the end it is hard to say what the smartest way to travel is: with baggage or without? It depends where you are going. Checking your disciplinary baggage should never be equated with traveling light. It is always a bit of a risk, but in the end one is more often than not reunited with their suitcase, the contents of which have almost always shifted.

This volume checks its disciplinary baggage so it can fly unencumbered for a brief while. In doing so it generates new insights as it meets at the interstices of interdisciplinary research. It ultimately provokes the necessity for even more thinking about time. We might find that it no longer makes sense to have a resident time person in every field. Perhaps it makes more sense to flood the disciplinary terrain with temporal terms that cannot be confined. Instead they remain open to the rhythmic variation of other disciplinary approaches such that it is never about time or space, but where the geopolitical and the chronopolitical meet. This volume wonderfully enlivens and makes visible this crucial intersection.

References

Sharma, S. (2013) Critical Time. *Communication and Critical/Cultural Studies*, 10(2–3), pp. 312–318.

Sharma, S. (2014) *In the Meantime: Temporality and Cultural Politics*. Durham, NC and London: Duke University Press.

Index